S0-DKM-900

WITHDRAWN FROM
KENT STATE UNIVERSITY LIBRARIES

DEPOLARIZATION
AND RELATED RATIOS OF
LIGHT SCATTERING
BY SPHEROIDS

DEPOLARIZATION AND RELATED RATIOS OF LIGHT SCATTERING BY SPHEROIDS

by Wilfried Heller / Masayuki Nakagaki / Gary D. Langolf

Wayne State University and Kyoto University

Wayne State University Press/Detroit/1974

Copyright © 1974 by Wayne State University Press, Detroit, Michigan 48202.
All rights are reserved. No part of this book may be reproduced without formal permission.

Published simultaneously in Canada
by the Copp Clark Publishing Company,
517 Wellington Street, West
Toronto 2B, Canada

Library of Congress Cataloging in Publication Data

Heller, Wilfried, 1903-
   Depolarization and related ratios of light scattering by spheroids.

   Includes bibliographical references.
   1. Light—Scattering—Tables, etc. 2. Polarization
(Light)—Tables. 3. Spheroidal state—Optical prop-
perties—Tables. I. Nakagaki, Masayuki, 1923-
joint author. II. Langolf, Gary D., joint author.
III. Title.
QC427.6.H43   535'.4   74-13816
ISBN 0-8143-1527-5

Contents

For p = 1 see Tables indicated in reference

Introduction and Acknowledgments

The tables in this book give numerical values of the depolarization of light scattered by dielectric, internally isotropic, nonabsorbing bodies of nonspherical shape which can be approximated as spheroids. In addition to data on depolarization, which is determined at an angle of observation, θ, of 90° with respect to the incident primary beam, the tables give the variation, with θ of related ratios of light scattering. The data given are based upon light scattering functions published by the authors in 1972.[1] They are, therefore, again based upon a generalization[2] of a theory by Stevenson[3] which in turn represents an extension of the Rayleigh[4]-Gans[5] theory of light scattering by spheroids of any relative refractive index.[6] While the previously published functions make it possible to determine the dimensions of nonspherical scatterers from measurements of the intensity of light scattered either at $\theta=90°$[1] or as a function of θ[7] or from spectra of these quantities,[7] the data in the present book should allow one to obtain the same information from measurements of intensity ratios, at $\theta=90°$ (depolarization) or at other angles θ (scattering ratio). The primary practical interest in determining intensity ratios rather than intensities themselves resides in the fact that no apparatus calibrations are necessary, that the solid angle at which observations are made is far less critical,[8] and above all, that the measured ratios can be directly compared with the theoretical data given in this book without need for data conversion or calculations.

From the definitions given below, it follows that the data cover those spheroidal bodies whose largest dimension does not exceed approximately 1/3 of the wavelength of light (in the respective solvent or dispersing medium) used in the scattering experiments. Thus, assuming the use of the green Hg-line and water at 25° C as the solvent or dispersing medium, the limiting largest dimension of a spheroid is 1300 A. U. Expressing this in terms of molecular weight, assuming a compact macromolecule with a partial specific volume of 1.0, the upper limits would be as follows for the axial ratios p given:

p	$M \times 10^{-8}$
50	0.00277
10	0.0692
5	0.277
2	1.73
0.5	3.46
0.2	1.38
0.1	0.692

Therefore, all biologically important macromolecules are accessible unless p is in excess of 10.

The authors are indebted to the late Dr. Walter Hoffmann and to Dr. Charles F. Briggs of the Wayne State University Computing Center for making the facilities of the Center available and for assisting us in many ways. We also would like to acknowledge the very helpful assistance received in connection with computer operations by Mr. Curtis Brock, Miss Catherine Chafey and Mr. John Mazay. The authors also wish to express their most sincere thanks to Dr. Herbert M. Schueller, director, and Mr. Richard Kinney, production manager of the Wayne State University Press, for their interest and help in making these data available in book form. The authors also would like to acknowledge the support received from the Office of Naval Research and from the Japan Society for the Promotion of Science.

Wilfried Heller, Masayuki Nakagaki and Gary D. Langolf

Definitions and Technical Information

A spheroid is defined dimensionally by the axial ratio

$$p = a/b \tag{1}$$

where a is the semiaxis of symmetry and b is the length of the semi-transverse axis. A spheroid is prolate if $p > 1$ and oblate if $p < 1$. The limiting characteristic shapes are rods and disks, respectively. Since the scattering power of a body depends on its dimensions relative to the wavelength of the radiation scattered, the spheroid dimensions are defined by the two parameters p and

$$\alpha = 2\pi a/\lambda_1 = 2\pi a n_1/\lambda_0 \tag{2}$$

where λ_1 is the wavelength in the solvent (medium), λ_0 that in the vacuum and n_1 is the refractive index of the solvent (medium). In the case of those nonspherical macromolecules which can be treated as rigid, internally isotropic and homogeneous bodies, such as most globular proteins in the native state, the relation between α and the molecular weight M is simply

$$M = N_A \, (\lambda_0\alpha)^3/6\overline{V} \, (\pi p)^2 n_1{}^3 \tag{3}$$

where N_A is Avogadro's number and $\overline{V}$ is the partial specific volume of the macromolecule. One thus arrives at the upper limits of molecular weights covered by the present tables as given in the Introduction, assuming $\overline{V} = 1.0$ and the use of visible radiation for the scattering experiments.

The tables give three ratios of intensities of scattered light, ρ_σ, ρ_π and ρ_u which, for technical reasons, are written RHO S, RHO P, AND RHO U, respectively. They are defined as follows:

$$\rho_\sigma = i(\sigma, \pi)/i(\sigma, \sigma) \tag{4}$$
$$\rho_\pi = i(\pi, \sigma)/i(\pi, \pi) \tag{5}$$
$$\rho_u = i(u, \pi)/i(u, \sigma) \tag{6}$$

Here, i is the theoretical light scattering function which is related to actually measured quantities by the relation

$$i(\theta) = (2r\pi/\lambda)^2 \, J(\theta) \tag{7}$$

J is that part of the radiant energy scattered by a single particle (molecule) from an incident beam of unit intensity which passes, at the angle of observation, θ, with respect to the incident beam, per unit time and unit area through a differential surface element of a sphere of radius r. One may, alternately, relate $i(\theta)$ to $J'(\theta)$, viz.:

$$i(\theta) = (2\pi/\lambda)^2 \, J'(\theta) \tag{8}$$

where $J'(\theta)$ represents the intensity of radiation scattered by a single particle (molecule), at the angle θ, per unit solid angle and per unit intensity of the incident beam. Dimensionally, J' represents the differential angular scattering cross section of a particle (molecule). Considering a solution (dispersed system) containing N spheroidal bodies per cm^3, the total radiant energy scattered per unit time by its unit volume and per unit solid angle in the θ direction

$$E_\theta = NJ'_\theta \, I_0 \text{ ergs cm}^{-3} \text{ sec}^{-1} \tag{9}$$

where I_0 is the intensity of the incident beam in ergs cm^{-2} sec^{-1}. Using, instead of the number concentration, the weight concentration c' (g ml^{-1})

$$\frac{E_\theta}{I_0 \, c'} = \frac{3p^2 \, \overline{V}}{2\lambda\alpha^3} i_\theta \tag{10}$$

The definition of the symbols given in brackets in equation (4), (5) and (6) is as follows: σ stands for perpendicular and π for parallel in line with its usage in general spectroscopy. The first symbol in the parentheses indicates the orientation of the electric vector of an incident linearly polarized beam with respect to the plane of observation. In the case of ρ_u, the first symbol, u, indicates that the incident beam is unpolarized. The second symbol expresses the orientation, with respect to the plane of observation, of the plane of polarization of a polarizing prism which is interposed between scattering system and observer. The three entries defined by equations (4), (5) and (6) take care of all possible experimental permutations—except for observation of the total scattered intensities, a case treated below—since Krishnan's reciprocity theorem[9]

$$i(\sigma \, \pi) = i(\pi \, \sigma) \tag{11}$$

is valid in a system containing randomly oriented spheroids.

Of particular practical interest are the values of ρ_σ, ρ_π and ρ_u if $\theta = 90°$. They then represent depolarizations, quantities which traditionally have, in various forms, been used as a measure of the form and/or intrinsic anisotropy of particles (molecules) much smaller than those that, on the basis of the present tables, have now become accessible to quantitative treatment. Several other degrees of depolarization, in addition to, or instead of, those given in equations (4), (5) and (6) are in use. The most important one is

$$\rho_t = \frac{i(\pi, t)}{i(\sigma, t)} \tag{12}$$

Here, the symbol t indicates that the total intensity of the light scattered at $\theta = 90°$ is measured. In resting systems, $\rho_t = \rho_u$, so that this degree of depolarization is covered implicitly by the tables. Another degree of depolarization

$$\rho_\rho = [i(\pi, \pi)/i(\sigma, \sigma)] = \rho_\sigma/\rho_\pi \tag{13}$$

can easily be calculated from the data given. The same applies to

$$\rho'_\pi = \frac{i(\pi, \pi)}{i(\pi, \sigma)} = 1/\rho_\pi \tag{14}$$

If any of these six ratios is determined at an angle $\theta \neq 90°$, it can no longer be called depolarization since even in the case of infinitely small spherical isotropic particles its value would deviate from zero. The quantity $\rho_t(\theta)$ is then defined as scattering ratio[8, 10] and the quantity $\rho_u(\theta)$ is defined as polarization ratio.[11] There exists no literature on $\rho_\sigma(\theta)$ or $\rho_\pi(\theta)$ of spheroids of finite size except for the theoretical data given in the present book.

The depolarizations, scattering ratios and polarization ratios are given for a series of m-values between 1.05–1.40, Δm being sufficiently small so that interpolations should be easy and precise for any m-value that may present itself in practice. The quantity $m = n_2/n_1$ where n_2 is the refractive index of the scattering body. If the scattering body is a macromolecular solute, where n_2 would be difficult to evaluate directly, one will use the simple relation[12]

$$(dn/dc) = (3/2)\,\bar{V}n_1\,[(m^2 - 1)/(m^2 + 2)] \tag{15}$$

in order to correlate the quantity m with the value of the slope of the straight line obtained on plotting the variation of the refractive index of the solution, n_{12}, against the solute concentration c(mg/ml).

The present tables apply to monodisperse systems. They may, however, be applied also to heterodisperse systems, in which there is a distribution of particle sizes or molecular weight, i.e., of α while p (the shape) is constant or, alternately, to systems in which there is a distribution of p while α is constant. One may then simply use the theoretical data given here in conjunction with the distribution function given some time ago.[13] . Simultaneous determination of distributions of both α (M) and p in doubly heterodisperse systems is also possible by making use of several of the ratios listed and by determining both their variation with θ and with λ. Details will have to be discussed elsewhere.

Although the calculations are strictly correct for intrinsically isotropic bodies, there is little doubt that corrections to be applied in the case of intrinsic anisotropy would be inconsequential as long as the refractive index values n_e and n_o —assuming uniaxial systems would differ only in the third or still higher decimals.

The parameter values considered in the tables are

θ = 0.5, 10 (10) 40, 45, 50 (10) 130, 135 (15) 180°
p = 0.1, 0.2, 0.5, 2, 5, 10, 20, 50
$\alpha_{p<1}$ = 0.02 (0.02) 0.08
$\alpha_{p>1}$ = 0.1, 0.2 (0.2) 1.0
m = 1.05, 1.10 (0.10), 1.40

For technical reasons ρ_σ, ρ_π and ρ_u are listed in the tables as RHO S, RHO P and RHO U, respectively; θ and α are written out as (THETA and ALPHA), p is replaced by P and m by M.

The computations were carried out with the IBM 360 Model 65 using the Fortran IV programming system and floating point arithmetic of double precision. The number of significant figures obtained was, in the most unfavorable cases, five and in general appreciably larger. Five figures are given for all entries which, therefore, are all significant.

References

1. W. Heller, M. Nakagaki and G. D. Langolf, Angular Scattering Functions for Spheroids, Wayne State University Press, Detroit, 1972.
2. W. Heller and M. Nakagaki, J. Chem. Physics, (1974).
3. A. F. Stevenson, J. Applied Physics, *24,* 1134, 1143 (1953).
4. Lord Rayleigh, Phil. Mag. (5) *44,* 28 (1897); (6) *35,* 373 (1918).
5. R. Gans, Ann. Physik (4) *37,* 881 (1912); *62,* 331 (1920).
6. This Rayleigh-Gans theory is not to be confused with the alternate more widely known Rayleigh-Gans theory which deals with the scattering of bodies of any shape within a far wider size range than encompassed by the present theory but which is restricted by the requirement that the refractive index of the scattering body differs very little from that of the medium (solvent). A severe shortcoming of that theory, in terms of the present objective, is that it would yield, for all parameter values considered, a value of zero for depolarization, scattering ratio and polarization ratio.
7. To be published shortly elsewhere.
8. W. Heller and R. Tabibian, J. Phys. Chem., *66,* 2059 (1962).
9. R. S. Krishnan, Proc. Indian Acad. Sci. *A1* 782 (1934); *A7,* 21 (1938); A. F. Stevenson; J. Applied Physics, *28,* 1015 (1957).
10. W. Heller, W. J. Pangonis and N. A. Economou, J. Chem. Physics, *34,* 971 (1961).
11. D. Sinclair and V. K. La Mer, Chem. Reviews, *44,* 245 (1949); M. Kerker, J. Colloid Sci., *5,* 165 (1950).
12. W. Heller, J. Phys. Chem., *69,* 1123 (1965).
13. A. F. Stevenson, W. Heller and M. L. Wallach, J. Chem. Physics, *34,* 1789 (1961).

ALPHA = 0.10

THETA	RHO S	RHO P	RHO U
0.	0.16757D-03	0.16757D-03	0.10000D 01
5.	0.16757D-03	0.16885D-03	0.99240D 00
10.	0.16757D-03	0.17278D-03	0.96983D 00
20.	0.16758D-03	0.18979D-03	0.88297D 00
30.	0.16759D-03	0.22348D-03	0.74992D 00
40.	0.16760D-03	0.28568D-03	0.58675D 00
45.	0.16761D-03	0.33531D-03	0.49995D 00
50.	0.16762D-03	0.40580D-03	0.41316D 00
60.	0.16765D-03	0.67070D-03	0.25009D 00
70.	0.16768D-03	0.14328D-02	0.11718D 00
80.	0.16773D-03	0.55373D-02	0.30453D-01
90.	0.16778D-03	0.10006D 01	0.33540D-03
100.	0.16783D-03	0.55407D-02	0.30454D-01
110.	0.16789D-03	0.14345D-02	0.11718D 00
120.	0.16795D-03	0.67192D-03	0.25009D 00
130.	0.16801D-03	0.40675D-03	0.41316D 00
135.	0.16804D-03	0.33617D-03	0.49995D 00
150.	0.16811D-03	0.22419D-03	0.74992D 00
165.	0.16816D-03	0.18024D-03	0.93298D 00
180.	0.16818D-03	0.16818D-03	0.10000D 01

ALPHA = 0.20

THETA	RHO S	RHO P	RHO U
0.	0.16704D-03	0.16704D-03	0.10000D 01
5.	0.16704D-03	0.16832D-03	0.99238D 00
10.	0.16704D-03	0.17225D-03	0.96975D 00
20.	0.16706D-03	0.18927D-03	0.88270D 00
30.	0.16710D-03	0.22299D-03	0.74943D 00
40.	0.16716D-03	0.28524D-03	0.58612D 00
45.	0.16720D-03	0.33493D-03	0.49930D 00
50.	0.16725D-03	0.40551D-03	0.41253D 00
60.	0.16736D-03	0.67084D-03	0.24960D 00
70.	0.16750D-03	0.14345D-02	0.11691D 00
80.	0.16767D-03	0.55495D-02	0.30376D-01
90.	0.16787D-03	0.10023D 01	0.33530D-03
100.	0.16810D-03	0.55634D-02	0.30378D-01
110.	0.16834D-03	0.14416D-02	0.11691D 00
120.	0.16858D-03	0.67575D-03	0.24960D 00
130.	0.16882D-03	0.40933D-03	0.41253D 00
135.	0.16893D-03	0.33840D-03	0.49929D 00
150.	0.16923D-03	0.22582D-03	0.74943D 00
165.	0.16942D-03	0.18162D-03	0.93282D 00
180.	0.16949D-03	0.16949D-03	0.10000D 01

ALPHA = 0.40

THETA	RHO S	RHO P	RHO U
0.	0.16490D-03	0.16490D-03	0.10000D 01
5.	0.16491D-03	0.16619D-03	0.99230D 00
10.	0.16493D-03	0.17013D-03	0.96945D 00
20.	0.16501D-03	0.18718D-03	0.88162D 00
30.	0.16517D-03	0.22098D-03	0.74747D 00
40.	0.16540D-03	0.28346D-03	0.58357D 00
45.	0.16555D-03	0.33338D-03	0.49667D 00
50.	0.16573D-03	0.40434D-03	0.40997D 00
60.	0.16617D-03	0.67144D-03	0.24761D 00
70.	0.16674D-03	0.14416D-02	0.11581D 00
80.	0.16744D-03	0.56000D-02	0.30062D-01
90.	0.16825D-03	0.10094D 01	0.33488D-03
100.	0.16917D-03	0.56574D-02	0.30066D-01
110.	0.17015D-03	0.14712D-02	0.11581D 00
120.	0.17116D-03	0.69173D-03	0.24757D 00
130.	0.17215D-03	0.42010D-03	0.40990D 00
135.	0.17262D-03	0.34769D-03	0.49658D 00
150.	0.17384D-03	0.23262D-03	0.74738D 00
165.	0.17466D-03	0.18737D-03	0.93213D 00
180.	0.17494D-03	0.17494D-03	0.10000D 01

ALPHA = 0.60

THETA	RHO S	RHO P	RHO U
0.	0.16135D-03	0.16135D-03	0.10000D 01
5.	0.16136D-03	0.16263D-03	0.99217D 00
10.	0.16140D-03	0.16658D-03	0.96894D 00
20.	0.16159D-03	0.18366D-03	0.87982D 00
30.	0.16192D-03	0.21759D-03	0.74418D 00
40.	0.16243D-03	0.28042D-03	0.57929D 00
45.	0.16276D-03	0.33072D-03	0.49223D 00
50.	0.16316D-03	0.40232D-03	0.40565D 00
60.	0.16415D-03	0.67247D-03	0.24423D 00
70.	0.16544D-03	0.14540D-02	0.11393D 00
80.	0.16703D-03	0.56896D-02	0.29519D-01
90.	0.16891D-03	0.10219D 01	0.33414D-03
100.	0.17104D-03	0.58271D-02	0.29519D-01
110.	0.17336D-03	0.15248D-02	0.11384D 00
120.	0.17576D-03	0.72081D-03	0.24397D 00
130.	0.17812D-03	0.43972D-03	0.40519D 00
135.	0.17925D-03	0.36462D-03	0.49170D 00
150.	0.18219D-03	0.24501D-03	0.74368D 00
165.	0.18417D-03	0.19785D-03	0.93089D 00
180.	0.18486D-03	0.18486D-03	0.10000D 01

ALPHA = 0.80

THETA	RHO S	RHO P	RHO U
0.	0.15636D-03	0.15636D-03	0.10000D 01
5.	0.15638D-03	0.15765D-03	0.99199D 00
10.	0.15646D-03	0.16159D-03	0.96823D 00
20.	0.15676D-03	0.17870D-03	0.87728D 00
30.	0.15732D-03	0.21275D-03	0.73952D 00
40.	0.15820D-03	0.27603D-03	0.57319D 00
45.	0.15878D-03	0.32685D-03	0.48589D 00
50.	0.15948D-03	0.39934D-03	0.39945D 00
60.	0.16124D-03	0.67401D-03	0.23934D 00
70.	0.16354D-03	0.14730D-02	0.11117D 00
80.	0.16643D-03	0.58285D-02	0.28716D-01
90.	0.16990D-03	0.10410D 01	0.33305D-03
100.	0.17389D-03	0.60966D-02	0.28691D-01
110.	0.17828D-03	0.16108D-02	0.11084D 00
120.	0.18289D-03	0.76765D-03	0.23838D 00
130.	0.18748D-03	0.47141D-03	0.39781D 00
135.	0.18968D-03	0.39197D-03	0.48401D 00
150.	0.19550D-03	0.26501D-03	0.73778D 00
165.	0.19945D-03	0.21472D-03	0.92889D 00
180.	0.20084D-03	0.20084D-03	0.10000D 01

ALPHA = 1.00

THETA	RHO S	RHO P	RHO U
0.	0.14994D-03	0.14994D-03	0.10000D 01
5.	0.14997D-03	0.15122D-03	0.99175D 00
10.	0.15008D-03	0.15515D-03	0.96730D 00
20.	0.15052D-03	0.17223D-03	0.87399D 00
30.	0.15134D-03	0.20635D-03	0.73346D 00
40.	0.15265D-03	0.27012D-03	0.56518D 00
45.	0.15353D-03	0.32157D-03	0.47752D 00
50.	0.15459D-03	0.39525D-03	0.39121D 00
60.	0.15731D-03	0.67619D-03	0.23276D 00
70.	0.16095D-03	0.15005D-02	0.10741D 00
80.	0.16560D-03	0.60345D-02	0.27603D-01
90.	0.17129D-03	0.10688D 01	0.33150D-03
100.	0.17797D-03	0.65121D-02	0.27503D-01
110.	0.18547D-03	0.17450D-02	0.10645D 00
120.	0.19348D-03	0.84148D-03	0.23008D 00
130.	0.20161D-03	0.52158D-03	0.38666D 00
135.	0.20556D-03	0.43531D-03	0.47232D 00
150.	0.21614D-03	0.29666D-03	0.72863D 00
165.	0.22342D-03	0.24135D-03	0.92576D 00
180.	0.22602D-03	0.22602D-03	0.10000D 01

ALPHA = 0.10

THETA	RHO S	RHO P	RHO U
0.	0.67872D-03	0.67872D-03	0.10000D 01
5.	0.67872D-03	0.68391D-03	0.99241D 00
10.	0.67872D-03	0.69983D-03	0.96986D 00
20.	0.67874D-03	0.76867D-03	0.88309D 00
30.	0.67878D-03	0.90502D-03	0.75018D 00
40.	0.67883D-03	0.11566D-02	0.58718D 00
45.	0.67887D-03	0.13574D-02	0.50047D 00
50.	0.67891D-03	0.16424D-02	0.41377D 00
60.	0.67901D-03	0.27123D-02	0.25086D 00
70.	0.67915D-03	0.57805D-02	0.11809D 00
80.	0.67931D-03	0.22065D-01	0.31445D-01
90.	0.67950D-03	0.10006D 01	0.13577D-02
100.	0.67972D-03	0.22077D-01	0.31447D-01
110.	0.67995D-03	0.57872D-02	0.11809D 00
120.	0.68019D-03	0.27169D-02	0.25086D 00
130.	0.68042D-03	0.16460D-02	0.41377D 00
135.	0.68053D-03	0.13607D-02	0.50047D 00
150.	0.68081D-03	0.90773D-03	0.75018D 00
165.	0.68100D-03	0.72990D-03	0.93305D 00
180.	0.68107D-03	0.68107D-03	0.10000D 01

ALPHA = 0.20

THETA	RHO S	RHO P	RHO U
0.	0.67670D-03	0.67670D-03	0.10000D 01
5.	0.67671D-03	0.68190D-03	0.99239D 00
10.	0.67673D-03	0.69782D-03	0.96979D 00
20.	0.67680D-03	0.76670D-03	0.88283D 00
30.	0.67694D-03	0.90315D-03	0.74970D 00
40.	0.67716D-03	0.11550D-02	0.58656D 00
45.	0.67730D-03	0.13560D-02	0.49982D 00
50.	0.67746D-03	0.16414D-02	0.41314D 00
60.	0.67788D-03	0.27130D-02	0.25037D 00
70.	0.67841D-03	0.57875D-02	0.11782D 00
80.	0.67907D-03	0.22113D-01	0.31367D-01
90.	0.67984D-03	0.10023D 01	0.13572D-02
100.	0.68071D-03	0.22162D-01	0.31374D-01
110.	0.68164D-03	0.58146D-02	0.11783D 00
120.	0.68260D-03	0.27318D-02	0.25039D 00
130.	0.68353D-03	0.16560D-02	0.41315D 00
135.	0.68397D-03	0.13693D-02	0.49983D 00
150.	0.68512D-03	0.91406D-03	0.74970D 00
165.	0.68588D-03	0.73526D-03	0.93289D 00
180.	0.68615D-03	0.68615D-03	0.10000D 01

ALPHA = 0.40

THETA	RHO S	RHO P	RHO U
0.	0.66864D-03	0.66864D-03	0.10000D 01
5.	0.66867D-03	0.67385D-03	0.99231D 00
10.	0.66874D-03	0.68980D-03	0.96949D 00
20.	0.66904D-03	0.75881D-03	0.88177D 00
30.	0.66958D-03	0.89563D-03	0.74778D 00
40.	0.67043D-03	0.11484D-02	0.58406D 00
45.	0.67099D-03	0.13503D-02	0.49724D 00
50.	0.67165D-03	0.16372D-02	0.41063D 00
60.	0.67331D-03	0.27159D-02	0.24842D 00
70.	0.67545D-03	0.58161D-02	0.11673D 00
80.	0.67811D-03	0.22310D-01	0.31051D-01
90.	0.68124D-03	0.10094D 01	0.13552D-02
100.	0.68477D-03	0.22515D-01	0.31077D-01
110.	0.68859D-03	0.59282D-02	0.11676D 00
120.	0.69253D-03	0.27935D-02	0.24843D 00
130.	0.69639D-03	0.16977D-02	0.41061D 00
135.	0.69822D-03	0.14053D-02	0.49720D 00
150.	0.70299D-03	0.94040D-03	0.74772D 00
165.	0.70618D-03	0.75756D-03	0.93223D 00
180.	0.70730D-03	0.70730D-03	0.10000D 01

ALPHA = 0.60

THETA	RHO S	RHO P	RHO U
0.	0.65519D-03	0.65519D-03	0.10000D 01
5.	0.65524D-03	0.66041D-03	0.99218D 00
10.	0.65540D-03	0.67638D-03	0.96899D 00
20.	0.65604D-03	0.74557D-03	0.88000D 00
30.	0.65722D-03	0.88291D-03	0.74455D 00
40.	0.65909D-03	0.11372D-02	0.57985D 00
45.	0.66032D-03	0.13406D-02	0.49289D 00
50.	0.66180D-03	0.16301D-02	0.40638D 00
60.	0.66552D-03	0.27209D-02	0.24509D 00
70.	0.67038D-03	0.58664D-02	0.11487D 00
80.	0.67644D-03	0.22661D-01	0.30506D-01
90.	0.68366D-03	0.10219D 01	0.13517D-02
100.	0.69189D-03	0.23151D-01	0.30557D-01
110.	0.70086D-03	0.61339D-02	0.11488D 00
120.	0.71018D-03	0.29055D-02	0.24496D 00
130.	0.71938D-03	0.17734D-02	0.40606D 00
135.	0.72376D-03	0.14707D-02	0.49249D 00
150.	0.73526D-03	0.98832D-03	0.74414D 00
165.	0.74298D-03	0.79807D-03	0.93103D 00
180.	0.74570D-03	0.74570D-03	0.10000D 01

ALPHA = 0.80

THETA	RHO S	RHO P	RHO U
0.	0.63632D-03	0.63632D-03	0.10000D 01
5.	0.63640D-03	0.64153D-03	0.99200D 00
10.	0.63666D-03	0.65752D-03	0.96829D 00
20.	0.63774D-03	0.72682D-03	0.87751D 00
30.	0.63974D-03	0.86474D-03	0.73998D 00
40.	0.64294D-03	0.11209D-02	0.57386D 00
45.	0.64509D-03	0.13265D-02	0.48666D 00
50.	0.64767D-03	0.16196D-02	0.40029D 00
60.	0.65425D-03	0.27285D-02	0.24028D 00
70.	0.66296D-03	0.59430D-02	0.11214D 00
80.	0.67398D-03	0.23204D-01	0.29700D-01
90.	0.68728D-03	0.10410D 01	0.13466D-02
100.	0.70266D-03	0.24156D-01	0.29770D-01
110.	0.71963D-03	0.64621D-02	0.11200D 00
120.	0.73749D-03	0.30854D-02	0.23959D 00
130.	0.75531D-03	0.18953D-02	0.39896D 00
135.	0.76387D-03	0.15760D-02	0.48509D 00
150.	0.78652D-03	0.10654D-02	0.73846D 00
165.	0.80188D-03	0.86312D-03	0.92910D 00
180.	0.80732D-03	0.80732D-03	0.10000D 01

ALPHA = 1.00

THETA	RHO S	RHO P	RHO U
0.	0.61197D-03	0.61197D-03	0.10000D 01
5.	0.61209D-03	0.61717D-03	0.99177D 00
10.	0.61246D-03	0.63312D-03	0.96738D 00
20.	0.61402D-03	0.70238D-03	0.87428D 00
30.	0.61695D-03	0.84070D-03	0.73402D 00
40.	0.62172D-03	0.10990D-02	0.56599D 00
45.	0.62497D-03	0.13072D-02	0.47843D 00
50.	0.62890D-03	0.16051D-02	0.39219D 00
60.	0.63909D-03	0.27392D-02	0.23380D 00
70.	0.65284D-03	0.60537D-02	0.10842D 00
80.	0.67056D-03	0.24008D-01	0.28583D-01
90.	0.69238D-03	0.10688D 01	0.13393D-02
100.	0.71808D-03	0.25695D-01	0.28643D-01
110.	0.74698D-03	0.69723D-02	0.10780D 00
120.	0.77793D-03	0.33672D-02	0.23163D 00
130.	0.80936D-03	0.20872D-02	0.38827D 00
135.	0.82464D-03	0.17417D-02	0.47389D 00
150.	0.86559D-03	0.11866D-02	0.72969D 00
165.	0.89377D-03	0.96516D-03	0.92610D 00
180.	0.90384D-03	0.90384D-03	0.10000D 01

ALPHA = 0.10

THETA	RHO S	RHO P	RHO U
0.	0.27624D-02	0.27624D-02	0.10000D 01
5.	0.27624D-02	0.27835D-02	0.99244D 00
10.	0.27624D-02	0.28481D-02	0.96999D 00
20.	0.27624D-02	0.31275D-02	0.88358D 00
30.	0.27625D-02	0.36808D-02	0.75122D 00
40.	0.27627D-02	0.47004D-02	0.58890D 00
45.	0.27628D-02	0.55127D-02	0.50255D 00
50.	0.27630D-02	0.66642D-02	0.41621D 00
60.	0.27633D-02	0.10969D-01	0.25398D 00
70.	0.27638D-02	0.23161D-01	0.12176D 00
80.	0.27644D-02	0.84260D-01	0.35474D-01
90.	0.27651D-02	0.10006D 01	0.55134D-02
100.	0.27659D-02	0.84294D-01	0.35481D-01
110.	0.27668D-02	0.23184D-01	0.12177D 00
120.	0.27677D-02	0.10986D-01	0.25399D 00
130.	0.27686D-02	0.66775D-02	0.41622D 00
135.	0.27690D-02	0.55249D-02	0.50256D 00
150.	0.27701D-02	0.36907D-02	0.75123D 00
165.	0.27708D-02	0.29693D-02	0.93333D 00
180.	0.27710D-02	0.27710D-02	0.10000D 01

ALPHA = 0.20

THETA	RHO S	RHO P	RHO U
0.	0.27552D-02	0.27552D-02	0.10000D 01
5.	0.27552D-02	0.27763D-02	0.99242D 00
10.	0.27552D-02	0.28409D-02	0.96992D 00
20.	0.27555D-02	0.31206D-02	0.88333D 00
30.	0.27559D-02	0.36742D-02	0.75076D 00
40.	0.27566D-02	0.46948D-02	0.58830D 00
45.	0.27571D-02	0.55081D-02	0.50193D 00
50.	0.27577D-02	0.66611D-02	0.41560D 00
60.	0.27591D-02	0.10973D-01	0.25350D 00
70.	0.27610D-02	0.23189D-01	0.12149D 00
80.	0.27634D-02	0.84440D-01	0.35392D-01
90.	0.27662D-02	0.10023D 01	0.55108D-02
100.	0.27694D-02	0.84573D-01	0.35417D-01
110.	0.27729D-02	0.23283D-01	0.12153D 00
120.	0.27765D-02	0.11041D-01	0.25356D 00
130.	0.27801D-02	0.67146D-02	0.41566D 00
135.	0.27818D-02	0.55570D-02	0.50198D 00
150.	0.27862D-02	0.37144D-02	0.75079D 00
165.	0.27891D-02	0.29894D-02	0.93319D 00
180.	0.27901D-02	0.27901D-02	0.10000D 01

ALPHA = 0.40

THETA	RHO S	RHO P	RHO U
0.	0.27264D-02	0.27264D-02	0.10000D 01
5.	0.27265D-02	0.27476D-02	0.99235D 00
10.	0.27267D-02	0.28123D-02	0.96963D 00
20.	0.27276D-02	0.30926D-02	0.88231D 00
30.	0.27293D-02	0.36477D-02	0.74890D 00
40.	0.27321D-02	0.46722D-02	0.58588D 00
45.	0.27339D-02	0.54892D-02	0.49942D 00
50.	0.27362D-02	0.66484D-02	0.41316D 00
60.	0.27420D-02	0.10988D-01	0.25159D 00
70.	0.27496D-02	0.23305D-01	0.12040D 00
80.	0.27592D-02	0.85177D-01	0.35056D-01
90.	0.27707D-02	0.10094D 01	0.55003D-02
100.	0.27838D-02	0.85730D-01	0.35158D-01
110.	0.27981D-02	0.23690D-01	0.12057D 00
120.	0.28130D-02	0.11267D-01	0.25177D 00
130.	0.28276D-02	0.68688D-02	0.41332D 00
135.	0.28346D-02	0.56903D-02	0.49956D 00
150.	0.28528D-02	0.38126D-02	0.74896D 00
165.	0.28650D-02	0.30727D-02	0.93257D 00
180.	0.28692D-02	0.28692D-02	0.10000D 01

ALPHA = 0.60

THETA	RHO S	RHO P	RHO U
0.	0.26783D-02	0.26783D-02	0.10000D 01
5.	0.26785D-02	0.26995D-02	0.99222D 00
10.	0.26789D-02	0.27645D-02	0.96915D 00
20.	0.26809D-02	0.30455D-02	0.88060D 00
30.	0.26846D-02	0.36030D-02	0.74578D 00
40.	0.26907D-02	0.46336D-02	0.58180D 00
45.	0.26948D-02	0.54569D-02	0.49519D 00
50.	0.26998D-02	0.66264D-02	0.40903D 00
60.	0.27127D-02	0.11015D-01	0.24832D 00
70.	0.27300D-02	0.23508D-01	0.11854D 00
80.	0.27520D-02	0.86486D-01	0.34477D-01
90.	0.27784D-02	0.10219D 01	0.54821D-02
100.	0.28090D-02	0.87800D-01	0.34704D-01
110.	0.28425D-02	0.24425D-01	0.11888D 00
120.	0.28776D-02	0.11676D-01	0.24861D 00
130.	0.29123D-02	0.71479D-02	0.40916D 00
135.	0.29289D-02	0.59320D-02	0.49523D 00
150.	0.29726D-02	0.39906D-02	0.74566D 00
165.	0.30020D-02	0.32236D-02	0.93146D 00
180.	0.30124D-02	0.30124D-02	0.10000D 01

P = 50.00   M = 1.20

ALPHA = 0.80

THETA	RHO S	RHO P	RHO U
0.	0.26107D-02	0.26107D-02	0.10000D 01
5.	0.26110D-02	0.26320D-02	0.99205D 00
10.	0.26117D-02	0.26970D-02	0.96847D 00
20.	0.26150D-02	0.29788D-02	0.87819D 00
30.	0.26213D-02	0.35390D-02	0.74136D 00
40.	0.26317D-02	0.45778D-02	0.57599D 00
45.	0.26389D-02	0.54098D-02	0.48914D 00
50.	0.26476D-02	0.65941D-02	0.40309D 00
60.	0.26705D-02	0.11054D-01	0.24360D 00
70.	0.27015D-02	0.23816D-01	0.11582D 00
80.	0.27413D-02	0.88507D-01	0.33622D-01
90.	0.27900D-02	0.10410D 01	0.54549D-02
100.	0.28469D-02	0.91041D-01	0.34020D-01
110.	0.29101D-02	0.25589D-01	0.11630D 00
120.	0.29770D-02	0.12328D-01	0.24374D 00
130.	0.30440D-02	0.75935D-02	0.40268D 00
135.	0.30762D-02	0.63179D-02	0.48848D 00
150.	0.31617D-02	0.42746D-02	0.74046D 00
165.	0.32197D-02	0.34640D-02	0.92970D 00
180.	0.32403D-02	0.32403D-02	0.10000D 01

ALPHA = 1.00

THETA	RHO S	RHO P	RHO U
0.	0.25233D-02	0.25233D-02	0.10000D 01
5.	0.25237D-02	0.25445D-02	0.99182D 00
10.	0.25247D-02	0.26095D-02	0.96759D 00
20.	0.25294D-02	0.28916D-02	0.87506D 00
30.	0.25385D-02	0.34542D-02	0.73558D 00
40.	0.25540D-02	0.45025D-02	0.56834D 00
45.	0.25649D-02	0.53456D-02	0.48113D 00
50.	0.25782D-02	0.65497D-02	0.39520D 00
60.	0.26136D-02	0.11110D-01	0.23724D 00
70.	0.26625D-02	0.24261D-01	0.11211D 00
80.	0.27265D-02	0.91490D-01	0.32440D-01
90.	0.28063D-02	0.10688D 01	0.54168D-02
100.	0.29010D-02	0.95933D-01	0.33045D-01
110.	0.30081D-02	0.27374D-01	0.11256D 00
120.	0.31233D-02	0.13336D-01	0.23658D 00
130.	0.32405D-02	0.82864D-02	0.39303D 00
135.	0.32975D-02	0.69185D-02	0.47834D 00
150.	0.34505D-02	0.47163D-02	0.73253D 00
165.	0.35558D-02	0.38369D-02	0.92699D 00
180.	0.35934D-02	0.35934D-02	0.10000D 01

P = 50.00   M = 1.30

ALPHA = 0.10

THETA	RHO S	RHO P	RHO U
0.	0.62642D-02	0.62642D-02	0.10000D 01
5.	0.62642D-02	0.63119D-02	0.99249D 00
10.	0.62642D-02	0.64579D-02	0.97020D 00
20.	0.62643D-02	0.70889D-02	0.88440D 00
30.	0.62645D-02	0.83369D-02	0.75297D 00
40.	0.62648D-02	0.10632D-01	0.59178D 00
45.	0.62650D-02	0.12457D-01	0.50603D 00
50.	0.62653D-02	0.15037D-01	0.42029D 00
60.	0.62660D-02	0.24616D-01	0.25919D 00
70.	0.62669D-02	0.51190D-01	0.12789D 00
80.	0.62681D-02	0.17314D 00	0.42207D-01
90.	0.62696D-02	0.10006D 01	0.12457D-01
100.	0.62712D-02	0.17317D 00	0.42220D-01
110.	0.62731D-02	0.51233D-01	0.12791D 00
120.	0.62750D-02	0.24649D-01	0.25922D 00
130.	0.62768D-02	0.15064D-01	0.42032D 00
135.	0.62777D-02	0.12482D-01	0.50605D 00
150.	0.62800D-02	0.83575D-02	0.75298D 00
165.	0.62816D-02	0.67299D-02	0.93380D 00
180.	0.62821D-02	0.62821D-02	0.10000D 01

ALPHA = 0.20

THETA	RHO S	RHO P	RHO U
0.	0.62498D-02	0.62498D-02	0.10000D 01
5.	0.62498D-02	0.62975D-02	0.99247D 00
10.	0.62499D-02	0.64436D-02	0.97013D 00
20.	0.62503D-02	0.70750D-02	0.88415D 00
30.	0.62510D-02	0.83239D-02	0.75252D 00
40.	0.62522D-02	0.10622D-01	0.59119D 00
45.	0.62531D-02	0.12448D-01	0.50542D 00
50.	0.62542D-02	0.15032D-01	0.41970D 00
60.	0.62570D-02	0.24626D-01	0.25872D 00
70.	0.62607D-02	0.51253D-01	0.12761D 00
80.	0.62655D-02	0.17350D 00	0.42115D-01
90.	0.62714D-02	0.10023D 01	0.12450D-01
100.	0.62781D-02	0.17365D 00	0.42166D-01
110.	0.62854D-02	0.51426D-01	0.12771D 00
120.	0.62931D-02	0.24760D-01	0.25883D 00
130.	0.63006D-02	0.15140D-01	0.41981D 00
135.	0.63042D-02	0.12548D-01	0.50552D 00
150.	0.63136D-02	0.84067D-02	0.75258D 00
165.	0.63199D-02	0.67719D-02	0.93367D 00
180.	0.63221D-02	0.63221D-02	0.10000D 01

10

ALPHA = 0.40

THETA	RHO S	RHO P	RHO U
0.	0.61921D-02	0.61921D-02	0.10000D 01
5.	0.61922D-02	0.62399D-02	0.99240D 00
10.	0.61926D-02	0.63863D-02	0.96985D 00
20.	0.61940D-02	0.70192D-02	0.88316D 00
30.	0.61969D-02	0.82717D-02	0.75071D 00
40.	0.62018D-02	0.10578D-01	0.58883D 00
45.	0.62052D-02	0.12413D-01	0.50297D 00
50.	0.62095D-02	0.15010D-01	0.41730D 00
60.	0.62206D-02	0.24666D-01	0.25681D 00
70.	0.62357D-02	0.51511D-01	0.12650D 00
80.	0.62551D-02	0.17499D 00	0.41739D-01
90.	0.62787D-02	0.10094D 01	0.12421D-01
100.	0.63060D-02	0.17563D 00	0.41946D-01
110.	0.63361D-02	0.52222D-01	0.12686D 00
120.	0.63674D-02	0.25217D-01	0.25724D 00
130.	0.63985D-02	0.15456D-01	0.41770D 00
135.	0.64133D-02	0.12823D-01	0.50333D 00
150.	0.64522D-02	0.86109D-02	0.75091D 00
165.	0.64783D-02	0.69459D-02	0.93311D 00
180.	0.64875D-02	0.64875D-02	0.10000D 01

ALPHA = 0.60

THETA	RHO S	RHO P	RHO U
0.	0.60956D-02	0.60956D-02	0.10000D 01
5.	0.60959D-02	0.61436D-02	0.99228D 00
10.	0.60966D-02	0.62903D-02	0.96938D 00
20.	0.60997D-02	0.69253D-02	0.88151D 00
30.	0.61059D-02	0.81833D-02	0.74768D 00
40.	0.61167D-02	0.10504D-01	0.58486D 00
45.	0.61243D-02	0.12353D-01	0.49883D 00
50.	0.61337D-02	0.14973D-01	0.41324D 00
60.	0.61586D-02	0.24737D-01	0.25356D 00
70.	0.61928D-02	0.51963D-01	0.12460D 00
80.	0.62372D-02	0.17764D 00	0.41092D-01
90.	0.62915D-02	0.10219D 01	0.12370D-01
100.	0.63549D-02	0.17916D 00	0.41562D-01
110.	0.64250D-02	0.53648D-01	0.12538D 00
120.	0.64989D-02	0.26040D-01	0.25442D 00
130.	0.65724D-02	0.16026D-01	0.41396D 00
135.	0.66076D-02	0.13318D-01	0.49944D 00
150.	0.67005D-02	0.89791D-02	0.74793D 00
165.	0.67632D-02	0.72594D-02	0.93210D 00
180.	0.67853D-02	0.67853D-02	0.10000D 01

ALPHA = 0.80

THETA	RHO S	RHO P	RHO U
0.	0.59597D-02	0.59597D-02	0.10000D 01
5.	0.59601D-02	0.60078D-02	0.99211D 00
10.	0.59612D-02	0.61549D-02	0.96873D 00
20.	0.59664D-02	0.67919D-02	0.87917D 00
30.	0.59768D-02	0.80567D-02	0.74338D 00
40.	0.59952D-02	0.10396D-01	0.57918D 00
45.	0.60084D-02	0.12266D-01	0.49291D 00
50.	0.60248D-02	0.14919D-01	0.40741D 00
60.	0.60689D-02	0.24841D-01	0.24887D 00
70.	0.61302D-02	0.52649D-01	0.12182D 00
80.	0.62107D-02	0.18172D 00	0.40138D-01
90.	0.63106D-02	0.10410D 01	0.12295D-01
100.	0.64283D-02	0.18462D 00	0.40985D-01
110.	0.65602D-02	0.55888D-01	0.12313D 00
120.	0.67004D-02	0.27339D-01	0.25011D 00
130.	0.68414D-02	0.16929D-01	0.40818D 00
135.	0.69094D-02	0.14104D-01	0.49339D 00
150.	0.70900D-02	0.95627D-02	0.74325D 00
165.	0.72129D-02	0.77557D-02	0.93051D 00
180.	0.72565D-02	0.72565D-02	0.10000D 01

ALPHA = 1.00

THETA	RHO S	RHO P	RHO U
0.	0.57836D-02	0.57836D-02	0.10000D 01
5.	0.57841D-02	0.58316D-02	0.99189D 00
10.	0.57856D-02	0.59788D-02	0.96787D 00
20.	0.57928D-02	0.66173D-02	0.87612D 00
30.	0.58078D-02	0.78887D-02	0.73774D 00
40.	0.58351D-02	0.10251D-01	0.57170D 00
45.	0.58550D-02	0.12146D-01	0.48506D 00
50.	0.58800D-02	0.14844D-01	0.39965D 00
60.	0.59483D-02	0.24989D-01	0.24254D 00
70.	0.60450D-02	0.53638D-01	0.11803D 00
80.	0.61742D-02	0.18774D 00	0.38822D-01
90.	0.63372D-02	0.10688D 01	0.12190D-01
100.	0.65326D-02	0.19273D 00	0.40165D-01
110.	0.67549D-02	0.59281D-01	0.11989D 00
120.	0.69949D-02	0.29327D-01	0.24381D 00
130.	0.72397D-02	0.18316D-01	0.39962D 00
135.	0.73590D-02	0.15312D-01	0.48439D 00
150.	0.76792D-02	0.10460D-01	0.73618D 00
165.	0.78997D-02	0.85170D-02	0.92809D 00
180.	0.79785D-02	0.79785D-02	0.10000D 01

ALPHA = 0.10

THETA	RHO S	RHO P	RHO U
0.	0.11127D-01	0.11127D-01	0.10000D 01
5.	0.11127D-01	0.11212D-01	0.99257D 00
10.	0.11128D-01	0.11470D-01	0.97049D 00
20.	0.11128D-01	0.12584D-01	0.88552D 00
30.	0.11128D-01	0.14785D-01	0.75536D 00
40.	0.11128D-01	0.18822D-01	0.59573D 00
45.	0.11129D-01	0.22021D-01	0.51081D 00
50.	0.11129D-01	0.26528D-01	0.42591D 00
60.	0.11130D-01	0.43107D-01	0.26636D 00
70.	0.11131D-01	0.87845D-01	0.13633D 00
80.	0.11133D-01	0.27208D 00	0.51479D-01
90.	0.11136D-01	0.10006D 01	0.22020D-01
100.	0.11138D-01	0.27211D 00	0.51499D-01
110.	0.11141D-01	0.87904D-01	0.13637D 00
120.	0.11145D-01	0.43158D-01	0.26641D 00
130.	0.11148D-01	0.26571D-01	0.42595D 00
135.	0.11149D-01	0.22060D-01	0.51086D 00
150.	0.11153D-01	0.14818D-01	0.75539D 00
165.	0.11156D-01	0.11948D-01	0.93445D 00
180.	0.11157D-01	0.11157D-01	0.10000D 01

ALPHA = 0.20

THETA	RHO S	RHO P	RHO U
0.	0.11105D-01	0.11105D-01	0.10000D 01
5.	0.11105D-01	0.11189D-01	0.99255D 00
10.	0.11105D-01	0.11447D-01	0.97042D 00
20.	0.11105D-01	0.12562D-01	0.88528D 00
30.	0.11106D-01	0.14765D-01	0.75492D 00
40.	0.11108D-01	0.18806D-01	0.59516D 00
45.	0.11109D-01	0.22008D-01	0.51022D 00
50.	0.11111D-01	0.26521D-01	0.42532D 00
60.	0.11115D-01	0.43126D-01	0.26589D 00
70.	0.11121D-01	0.87954D-01	0.13605D 00
80.	0.11128D-01	0.27265D 00	0.51372D-01
90.	0.11138D-01	0.10023D 01	0.22005D-01
100.	0.11149D-01	0.27273D 00	0.51453D-01
110.	0.11161D-01	0.88193D-01	0.13619D 00
120.	0.11174D-01	0.43331D-01	0.26607D 00
130.	0.11186D-01	0.26693D-01	0.42550D 00
135.	0.11192D-01	0.22167D-01	0.51038D 00
150.	0.11208D-01	0.14899D-01	0.75502D 00
165.	0.11219D-01	0.12017D-01	0.93433D 00
180.	0.11222D-01	0.11222D-01	0.10000D 01

ALPHA = 0.40

THETA	RHO S	RHO P	RHO U
0.	0.11014D-01	0.11014D-01	0.10000D 01
5.	0.11014D-01	0.11098D-01	0.99248D 00
10.	0.11014D-01	0.11357D-01	0.97015D 00
20.	0.11016D-01	0.12475D-01	0.88432D 00
30.	0.11019D-01	0.14684D-01	0.75317D 00
40.	0.11026D-01	0.18740D-01	0.59285D 00
45.	0.11031D-01	0.21957D-01	0.50781D 00
50.	0.11037D-01	0.26494D-01	0.42295D 00
60.	0.11054D-01	0.43205D-01	0.26397D 00
70.	0.11077D-01	0.88399D-01	0.13489D 00
80.	0.11108D-01	0.27496D 00	0.50939D-01
90.	0.11146D-01	0.10094D 01	0.21944D-01
100.	0.11191D-01	0.27530D 00	0.51265D-01
110.	0.11240D-01	0.89385D-01	0.13546D 00
120.	0.11292D-01	0.44047D-01	0.26467D 00
130.	0.11344D-01	0.27198D-01	0.42363D 00
135.	0.11369D-01	0.22609D-01	0.50843D 00
150.	0.11434D-01	0.15231D-01	0.75353D 00
165.	0.11478D-01	0.12301D-01	0.93382D 00
180.	0.11493D-01	0.11493D-01	0.10000D 01

ALPHA = 0.60

THETA	RHO S	RHO P	RHO U
0.	0.10861D-01	0.10861D-01	0.10000D 01
5.	0.10861D-01	0.10946D-01	0.99236D 00
10.	0.10862D-01	0.11205D-01	0.96970D 00
20.	0.10866D-01	0.12327D-01	0.88272D 00
30.	0.10873D-01	0.14546D-01	0.75021D 00
40.	0.10888D-01	0.18628D-01	0.58896D 00
45.	0.10898D-01	0.21869D-01	0.50374D 00
50.	0.10912D-01	0.26447D-01	0.41894D 00
60.	0.10949D-01	0.43343D-01	0.26071D 00
70.	0.11002D-01	0.89181D-01	0.13291D 00
80.	0.11072D-01	0.27907D 00	0.50193D-01
90.	0.11160D-01	0.10219D 01	0.21837D-01
100.	0.11263D-01	0.27985D 00	0.50938D-01
110.	0.11379D-01	0.91512D-01	0.13419D 00
120.	0.11501D-01	0.45329D-01	0.26221D 00
130.	0.11623D-01	0.28104D-01	0.42032D 00
135.	0.11682D-01	0.23402D-01	0.50498D 00
150.	0.11838D-01	0.15827D-01	0.75086D 00
165.	0.11943D-01	0.12812D-01	0.93292D 00
180.	0.11980D-01	0.11980D-01	0.10000D 01

ALPHA = 0.80

THETA	RHO S	RHO P	RHO U
0.	0.10646D-01	0.10646D-01	0.10000D 01
5.	0.10646D-01	0.10731D-01	0.99220D 00
10.	0.10648D-01	0.10991D-01	0.96906D 00
20.	0.10653D-01	0.12117D-01	0.88044D 00
30.	0.10666D-01	0.14349D-01	0.74602D 00
40.	0.10690D-01	0.18465D-01	0.58340D 00
45.	0.10709D-01	0.21742D-01	0.49791D 00
50.	0.10732D-01	0.26377D-01	0.41317D 00
60.	0.10798D-01	0.43548D-01	0.25599D 00
70.	0.10893D-01	0.90366D-01	0.13002D 00
80.	0.11020D-01	0.28538D 00	0.49096D-01
90.	0.11181D-01	0.10410D 01	0.21680D-01
100.	0.11373D-01	0.28683D 00	0.50448D-01
110.	0.11589D-01	0.94823D-01	0.13227D 00
120.	0.11820D-01	0.47339D-01	0.25845D 00
130.	0.12053D-01	0.29529D-01	0.41524D 00
135.	0.12166D-01	0.24650D-01	0.49964D 00
150.	0.12467D-01	0.16766D-01	0.74671D 00
165.	0.12671D-01	0.13616D-01	0.93151D 00
180.	0.12744D-01	0.12744D-01	0.10000D 01

ALPHA = 1.00

THETA	RHO S	RHO P	RHO U
0.	0.10367D-01	0.10367D-01	0.10000D 01
5.	0.10367D-01	0.10452D-01	0.99198D 00
10.	0.10368D-01	0.10712D-01	0.96823D 00
20.	0.10376D-01	0.11842D-01	0.87747D 00
30.	0.10393D-01	0.14087D-01	0.74051D 00
40.	0.10429D-01	0.18245D-01	0.57606D 00
45.	0.10457D-01	0.21567D-01	0.49019D 00
50.	0.10493D-01	0.26282D-01	0.40550D 00
60.	0.10595D-01	0.43837D-01	0.24965D 00
70.	0.10745D-01	0.92070D-01	0.12609D 00
80.	0.10949D-01	0.29467D 00	0.47584D-01
90.	0.11210D-01	0.10687D 01	0.21460D-01
100.	0.11527D-01	0.29709D 00	0.49754D-01
110.	0.11890D-01	0.99776D-01	0.12952D 00
120.	0.12283D-01	0.50377D-01	0.25300D 00
130.	0.12686D-01	0.31694D-01	0.40777D 00
135.	0.12882D-01	0.26549D-01	0.49177D 00
150.	0.13409D-01	0.18194D-01	0.74050D 00
165.	0.13773D-01	0.14835D-01	0.92938D 00
180.	0.13903D-01	0.13903D-01	0.10000D 01

ALPHA = 0.10

THETA	RHO S	RHO P	RHO U
0.	0.16217D-03	0.16217D-03	0.10000D 01
5.	0.16217D-03	0.16341D-03	0.99240D 00
10.	0.16217D-03	0.16722D-03	0.96983D 00
20.	0.16218D-03	0.18368D-03	0.88297D 00
30.	0.16219D-03	0.21629D-03	0.74992D 00
40.	0.16220D-03	0.27647D-03	0.58675D 00
45.	0.16221D-03	0.32451D-03	0.49995D 00
50.	0.16222D-03	0.39273D-03	0.41316D 00
60.	0.16225D-03	0.64911D-03	0.25008D 00
70.	0.16229D-03	0.13867D-02	0.11717D 00
80.	0.16233D-03	0.53599D-02	0.30443D-01
90.	0.16238D-03	0.10006D 01	0.32461D-03
100.	0.16243D-03	0.53633D-02	0.30443D-01
110.	0.16249D-03	0.13884D-02	0.11717D 00
120.	0.16255D-03	0.65030D-03	0.25008D 00
130.	0.16261D-03	0.39366D-03	0.41316D 00
135.	0.16263D-03	0.32536D-03	0.49995D 00
150.	0.16271D-03	0.21698D-03	0.74992D 00
165.	0.16275D-03	0.17445D-03	0.93298D 00
180.	0.16277D-03	0.16277D-03	0.10000D 01

ALPHA = 0.20

THETA	RHO S	RHO P	RHO U
0.	0.16166D-03	0.16166D-03	0.10000D 01
5.	0.16166D-03	0.16291D-03	0.99238D 00
10.	0.16167D-03	0.16671D-03	0.96975D 00
20.	0.16169D-03	0.18318D-03	0.88270D 00
30.	0.16173D-03	0.21581D-03	0.74943D 00
40.	0.16179D-03	0.27607D-03	0.58612D 00
45.	0.16183D-03	0.32416D-03	0.49929D 00
50.	0.16187D-03	0.39248D-03	0.41252D 00
60.	0.16198D-03	0.64930D-03	0.24959D 00
70.	0.16212D-03	0.13885D-02	0.11690D 00
80.	0.16229D-03	0.53722D-02	0.30366D-01
90.	0.16249D-03	0.10023D 01	0.32455D-03
100.	0.16271D-03	0.53860D-02	0.30367D-01
110.	0.16294D-03	0.13955D-02	0.11690D 00
120.	0.16318D-03	0.65412D-03	0.24959D 00
130.	0.16341D-03	0.39623D-03	0.41252D 00
135.	0.16352D-03	0.32757D-03	0.49929D 00
150.	0.16381D-03	0.21859D-03	0.74943D 00
165.	0.16400D-03	0.17581D-03	0.93282D 00
180.	0.16407D-03	0.16407D-03	0.10000D 01

ALPHA = 0.40

THETA	RHO S	RHO P	RHO U
0.	0.15963D-03	0.15963D-03	0.10000D 01
5.	0.15963D-03	0.16087D-03	0.99230D 00
10.	0.15965D-03	0.16469D-03	0.96945D 00
20.	0.15974D-03	0.18119D-03	0.88163D 00
30.	0.15989D-03	0.21392D-03	0.74747D 00
40.	0.16012D-03	0.27441D-03	0.58358D 00
45.	0.16027D-03	0.32274D-03	0.49667D 00
50.	0.16045D-03	0.39145D-03	0.40998D 00
60.	0.16089D-03	0.65007D-03	0.24761D 00
70.	0.16145D-03	0.13958D-02	0.11581D 00
80.	0.16213D-03	0.54230D-02	0.30054D-01
90.	0.16293D-03	0.10094D 01	0.32430D-03
100.	0.16383D-03	0.54801D-02	0.30054D-01
110.	0.16479D-03	0.14250D-02	0.11579D 00
120.	0.16578D-03	0.67000D-03	0.24756D 00
130.	0.16675D-03	0.40692D-03	0.40989D 00
135.	0.16721D-03	0.33678D-03	0.49657D 00
150.	0.16840D-03	0.22533D-03	0.74738D 00
165.	0.16919D-03	0.18151D-03	0.93213D 00
180.	0.16947D-03	0.16947D-03	0.10000D 01

ALPHA = 0.60

THETA	RHO S	RHO P	RHO U
0.	0.15623D-03	0.15623D-03	0.10000D 01
5.	0.15625D-03	0.15748D-03	0.99217D 00
10.	0.15629D-03	0.16130D-03	0.96894D 00
20.	0.15647D-03	0.17785D-03	0.87983D 00
30.	0.15680D-03	0.21072D-03	0.74419D 00
40.	0.15731D-03	0.27159D-03	0.57930D 00
45.	0.15765D-03	0.32032D-03	0.49225D 00
50.	0.15804D-03	0.38968D-03	0.40567D 00
60.	0.15903D-03	0.65142D-03	0.24425D 00
70.	0.16030D-03	0.14087D-02	0.11393D 00
80.	0.16186D-03	0.55134D-02	0.29515D-01
90.	0.16371D-03	0.10218D 01	0.32387D-03
100.	0.16580D-03	0.56500D-02	0.29506D-01
110.	0.16807D-03	0.14784D-02	0.11383D 00
120.	0.17042D-03	0.69892D-03	0.24396D 00
130.	0.17273D-03	0.42641D-03	0.40518D 00
135.	0.17383D-03	0.35359D-03	0.49170D 00
150.	0.17671D-03	0.23762D-03	0.74368D 00
165.	0.17863D-03	0.19190D-03	0.93089D 00
180.	0.17931D-03	0.17931D-03	0.10000D 01

ALPHA = 0.80

THETA	RHO S	RHO P	RHO U
0.	0.15148D-03	0.15148D-03	0.10000D 01
5.	0.15150D-03	0.15272D-03	0.99199D 00
10.	0.15157D-03	0.15655D-03	0.96823D 00
20.	0.15188D-03	0.17313D-03	0.87730D 00
30.	0.15244D-03	0.20614D-03	0.73955D 00
40.	0.15332D-03	0.26750D-03	0.57322D 00
45.	0.15390D-03	0.31677D-03	0.48593D 00
50.	0.15459D-03	0.38707D-03	0.39948D 00
60.	0.15634D-03	0.65343D-03	0.23937D 00
70.	0.15861D-03	0.14283D-02	0.11119D 00
80.	0.16146D-03	0.56532D-02	0.28718D-01
90.	0.16487D-03	0.10408D 01	0.32322D-03
100.	0.16879D-03	0.59199D-02	0.28676D-01
110.	0.17309D-03	0.15640D-02	0.11082D 00
120.	0.17761D-03	0.74550D-03	0.23837D 00
130.	0.18210D-03	0.45789D-03	0.39780D 00
135.	0.18425D-03	0.38076D-03	0.48401D 00
150.	0.18995D-03	0.25748D-03	0.73778D 00
165.	0.19380D-03	0.20864D-03	0.92889D 00
180.	0.19517D-03	0.19517D-03	0.10000D 01

ALPHA = 1.00

THETA	RHO S	RHO P	RHO U
0.	0.14535D-03	0.14535D-03	0.10000D 01
5.	0.14539D-03	0.14660D-03	0.99175D 00
10.	0.14549D-03	0.15041D-03	0.96731D 00
20.	0.14594D-03	0.16698D-03	0.87402D 00
30.	0.14677D-03	0.20010D-03	0.73351D 00
40.	0.14808D-03	0.26201D-03	0.56524D 00
45.	0.14896D-03	0.31195D-03	0.47759D 00
50.	0.15001D-03	0.38349D-03	0.39128D 00
60.	0.15272D-03	0.65628D-03	0.23282D 00
70.	0.15632D-03	0.14568D-02	0.10744D 00
80.	0.16090D-03	0.58605D-02	0.27612D-01
90.	0.16651D-03	0.10684D 01	0.32231D-03
100.	0.17308D-03	0.63362D-02	0.27484D-01
110.	0.18043D-03	0.16979D-02	0.10643D 00
120.	0.18829D-03	0.81897D-03	0.23006D 00
130.	0.19626D-03	0.50775D-03	0.38664D 00
135.	0.20013D-03	0.42381D-03	0.47231D 00
150.	0.21049D-03	0.28891D-03	0.72862D 00
165.	0.21762D-03	0.23508D-03	0.92576D 00
180.	0.22017D-03	0.22017D-03	0.10000D 01

ALPHA = 0.10

THETA	RHO S	RHO P	RHO U
0.	0.65651D-03	0.65651D-03	0.10000D 01
5.	0.65652D-03	0.66154D-03	0.99241D 00
10.	0.65652D-03	0.67694D-03	0.96986D 00
20.	0.65654D-03	0.74352D-03	0.88309D 00
30.	0.65657D-03	0.87543D-03	0.75017D 00
40.	0.65663D-03	0.11188D-02	0.58716D 00
45.	0.65666D-03	0.13130D-02	0.50044D 00
50.	0.65670D-03	0.15887D-02	0.41374D 00
60.	0.65681D-03	0.26237D-02	0.25083D 00
70.	0.65694D-03	0.55924D-02	0.11805D 00
80.	0.65710D-03	0.21358D-01	0.31402D-01
90.	0.65729D-03	0.10006D 01	0.13133D-02
100.	0.65750D-03	0.21371D-01	0.31403D-01
110.	0.65773D-03	0.55990D-02	0.11805D 00
120.	0.65796D-03	0.26283D-02	0.25083D 00
130.	0.65818D-03	0.15923D-02	0.41374D 00
135.	0.65829D-03	0.13163D-02	0.50045D 00
150.	0.65857D-03	0.87808D-03	0.75017D 00
165.	0.65875D-03	0.70606D-03	0.93305D 00
180.	0.65882D-03	0.65882D-03	0.10000D 01

ALPHA = 0.20

THETA	RHO S	RHO P	RHO U
0.	0.65460D-03	0.65460D-03	0.10000D 01
5.	0.65461D-03	0.65963D-03	0.99239D 00
10.	0.65463D-03	0.67504D-03	0.96979D 00
20.	0.65471D-03	0.74167D-03	0.88283D 00
30.	0.65484D-03	0.87368D-03	0.74969D 00
40.	0.65506D-03	0.11173D-02	0.58654D 00
45.	0.65520D-03	0.13118D-02	0.49981D 00
50.	0.65536D-03	0.15879D-02	0.41312D 00
60.	0.65578D-03	0.26247D-02	0.25034D 00
70.	0.65630D-03	0.55997D-02	0.11778D 00
80.	0.65695D-03	0.21407D-01	0.31325D-01
90.	0.65771D-03	0.10023D 01	0.13130D-02
100.	0.65856D-03	0.21457D-01	0.31330D-01
110.	0.65947D-03	0.56264D-02	0.11779D 00
120.	0.66040D-03	0.26431D-02	0.25035D 00
130.	0.66131D-03	0.16023D-02	0.41313D 00
135.	0.66174D-03	0.13249D-02	0.49981D 00
150.	0.66286D-03	0.88438D-03	0.74969D 00
165.	0.66361D-03	0.71138D-03	0.93289D 00
180.	0.66387D-03	0.66387D-03	0.10000D 01

ALPHA = 0.40

THETA	RHO S	RHO P	RHO U
0.	0.64696D-03	0.64696D-03	0.10000D 01
5.	0.64699D-03	0.65200D-03	0.99231D 00
10.	0.64706D-03	0.66744D-03	0.96949D 00
20.	0.64736D-03	0.73422D-03	0.88177D 00
30.	0.64790D-03	0.86663D-03	0.74777D 00
40.	0.64875D-03	0.11113D-02	0.58405D 00
45.	0.64931D-03	0.13067D-02	0.49724D 00
50.	0.64997D-03	0.15844D-02	0.41062D 00
60.	0.65161D-03	0.26284D-02	0.24840D 00
70.	0.65373D-03	0.56295D-02	0.11670D 00
80.	0.65634D-03	0.21606D-01	0.31014D-01
90.	0.65941D-03	0.10094D 01	0.13118D-02
100.	0.66288D-03	0.21813D-01	0.31032D-01
110.	0.66662D-03	0.57402D-02	0.11672D 00
120.	0.67047D-03	0.27047D-02	0.24839D 00
130.	0.67424D-03	0.16438D-02	0.41058D 00
135.	0.67603D-03	0.13607D-02	0.49718D 00
150.	0.68069D-03	0.91057D-03	0.74771D 00
165.	0.68379D-03	0.73354D-03	0.93223D 00
180.	0.68488D-03	0.68488D-03	0.10000D 01

ALPHA = 0.60

THETA	RHO S	RHO P	RHO U
0.	0.63421D-03	0.63421D-03	0.10000D 01
5.	0.63426D-03	0.63926D-03	0.99218D 00
10.	0.63442D-03	0.65474D-03	0.96899D 00
20.	0.63507D-03	0.72173D-03	0.88001D 00
30.	0.63626D-03	0.85474D-03	0.74456D 00
40.	0.63813D-03	0.11010D-02	0.57986D 00
45.	0.63936D-03	0.12980D-02	0.49289D 00
50.	0.64083D-03	0.15783D-02	0.40639D 00
60.	0.64452D-03	0.26349D-02	0.24509D 00
70.	0.64932D-03	0.56820D-02	0.11485D 00
80.	0.65529D-03	0.21960D-01	0.30475D-01
90.	0.66238D-03	0.10218D 01	0.13098D-02
100.	0.67045D-03	0.22455D-01	0.30508D-01
110.	0.67924D-03	0.59463D-02	0.11483D 00
120.	0.68836D-03	0.28166D-02	0.24491D 00
130.	0.69735D-03	0.17192D-02	0.40603D 00
135.	0.70164D-03	0.14258D-02	0.49246D 00
150.	0.71287D-03	0.95823D-03	0.74413D 00
165.	0.72040D-03	0.77381D-03	0.93102D 00
180.	0.72305D-03	0.72305D-03	0.10000D 01

ALPHA = 0.80

THETA	RHO S	RHO P	RHO U
0.	0.61633D-03	0.61633D-03	0.10000D 01
5.	0.61642D-03	0.62139D-03	0.99200D 00
10.	0.61668D-03	0.63688D-03	0.96829D 00
20.	0.61777D-03	0.70406D-03	0.87753D 00
30.	0.61980D-03	0.83774D-03	0.74001D 00
40.	0.62301D-03	0.10861D-02	0.57390D 00
45.	0.62516D-03	0.12853D-02	0.48670D 00
50.	0.62772D-03	0.15695D-02	0.40033D 00
60.	0.63425D-03	0.26447D-02	0.24030D 00
70.	0.64287D-03	0.57618D-02	0.11215D 00
80.	0.65373D-03	0.22507D-01	0.29680D-01
90.	0.66682D-03	0.10408D 01	0.13066D-02
100.	0.68191D-03	0.23470D-01	0.29717D-01
110.	0.69855D-03	0.62752D-02	0.11194D 00
120.	0.71604D-03	0.29961D-02	0.23953D 00
130.	0.73348D-03	0.18407D-02	0.39891D 00
135.	0.74186D-03	0.15306D-02	0.48505D 00
150.	0.76400D-03	0.10349D-02	0.73844D 00
165.	0.77900D-03	0.83850D-03	0.92910D 00
180.	0.78431D-03	0.78431D-03	0.10000D 01

ALPHA = 1.00

THETA	RHO S	RHO P	RHO U
0.	0.59328D-03	0.59328D-03	0.10000D 01
5.	0.59340D-03	0.59833D-03	0.99177D 00
10.	0.59378D-03	0.61381D-03	0.96739D 00
20.	0.59537D-03	0.68102D-03	0.87431D 00
30.	0.59834D-03	0.81527D-03	0.73408D 00
40.	0.60315D-03	0.10660D-02	0.56606D 00
45.	0.60640D-03	0.12681D-02	0.47851D 00
50.	0.61032D-03	0.15573D-02	0.39227D 00
60.	0.62046D-03	0.26585D-02	0.23386D 00
70.	0.63408D-03	0.58771D-02	0.10846D 00
80.	0.65158D-03	0.23316D-01	0.28578D-01
90.	0.67307D-03	0.10683D 01	0.13022D-02
100.	0.69832D-03	0.25026D-01	0.28583D-01
110.	0.72669D-03	0.67864D-02	0.10773D 00
120.	0.75705D-03	0.32776D-02	0.23156D 00
130.	0.78784D-03	0.20320D-02	0.38820D 00
135.	0.80281D-03	0.16958D-02	0.47382D 00
150.	0.84290D-03	0.11556D-02	0.72965D 00
165.	0.87049D-03	0.94003D-03	0.92609D 00
180.	0.88034D-03	0.88034D-03	0.10000D 01

ALPHA = 0.10

THETA	RHO S	RHO P	RHO U
0.	0.26693D-02	0.26693D-02	0.10000D 01
5.	0.26693D-02	0.26897D-02	0.99244D 00
10.	0.26694D-02	0.27522D-02	0.96998D 00
20.	0.26694D-02	0.30223D-02	0.88356D 00
30.	0.26695D-02	0.35569D-02	0.75118D 00
40.	0.26697D-02	0.45424D-02	0.58883D 00
45.	0.26698D-02	0.53276D-02	0.50246D 00
50.	0.26700D-02	0.64407D-02	0.41610D 00
60.	0.26703D-02	0.10603D-01	0.25384D 00
70.	0.26708D-02	0.22397D-01	0.12160D 00
80.	0.26714D-02	0.81648D-01	0.35295D-01
90.	0.26721D-02	0.10006D 01	0.53284D-02
100.	0.26729D-02	0.81683D-01	0.35301D-01
110.	0.26737D-02	0.22420D-01	0.12161D 00
120.	0.26746D-02	0.10619D-01	0.25385D 00
130.	0.26754D-02	0.64538D-02	0.41611D 00
135.	0.26758D-02	0.53395D-02	0.50247D 00
150.	0.26769D-02	0.35667D-02	0.75119D 00
165.	0.26776D-02	0.28695D-02	0.93332D 00
180.	0.26779D-02	0.26779D-02	0.10000D 01

ALPHA = 0.20

THETA	RHO S	RHO P	RHO U
0.	0.26626D-02	0.26626D-02	0.10000D 01
5.	0.26627D-02	0.26830D-02	0.99242D 00
10.	0.26627D-02	0.27455D-02	0.96991D 00
20.	0.26629D-02	0.30158D-02	0.88331D 00
30.	0.26634D-02	0.35509D-02	0.75072D 00
40.	0.26641D-02	0.45375D-02	0.58823D 00
45.	0.26646D-02	0.53237D-02	0.50184D 00
50.	0.26651D-02	0.64384D-02	0.41550D 00
60.	0.26666D-02	0.10608D-01	0.25337D 00
70.	0.26684D-02	0.22427D-01	0.12133D 00
80.	0.26708D-02	0.81830D-01	0.35215D-01
90.	0.26736D-02	0.10023D 01	0.53268D-02
100.	0.26767D-02	0.81971D-01	0.35237D-01
110.	0.26801D-02	0.22520D-01	0.12137D 00
120.	0.26837D-02	0.10674D-01	0.25342D 00
130.	0.26871D-02	0.64911D-02	0.41555D 00
135.	0.26888D-02	0.53717D-02	0.50188D 00
150.	0.26931D-02	0.35904D-02	0.75074D 00
165.	0.26959D-02	0.28895D-02	0.93317D 00
180.	0.26969D-02	0.26969D-02	0.10000D 01

ALPHA = 0.40

THETA	RHO S	RHO P	RHO U
0.	0.26358D-02	0.26358D-02	0.10000D 01
5.	0.26358D-02	0.26562D-02	0.99235D 00
10.	0.26361D-02	0.27189D-02	0.96962D 00
20.	0.26370D-02	0.29898D-02	0.88229D 00
30.	0.26387D-02	0.35267D-02	0.74887D 00
40.	0.26415D-02	0.45175D-02	0.58582D 00
45.	0.26434D-02	0.53077D-02	0.49935D 00
50.	0.26456D-02	0.64288D-02	0.41308D 00
60.	0.26514D-02	0.10627D-01	0.25147D 00
70.	0.26589D-02	0.22549D-01	0.12026D 00
80.	0.26684D-02	0.82577D-01	0.34889D-01
90.	0.26797D-02	0.10094D 01	0.53202D-02
100.	0.26926D-02	0.83163D-01	0.34975D-01
110.	0.27066D-02	0.22933D-01	0.12040D 00
120.	0.27211D-02	0.10902D-01	0.25162D 00
130.	0.27354D-02	0.66458D-02	0.41320D 00
135.	0.27422D-02	0.55055D-02	0.49945D 00
150.	0.27599D-02	0.36887D-02	0.74890D 00
165.	0.27718D-02	0.29729D-02	0.93256D 00
180.	0.27760D-02	0.27760D-02	0.10000D 01

ALPHA = 0.60

THETA	RHO S	RHO P	RHO U
0.	0.25909D-02	0.25909D-02	0.10000D 01
5.	0.25911D-02	0.26114D-02	0.99222D 00
10.	0.25915D-02	0.26743D-02	0.96915D 00
20.	0.25935D-02	0.29462D-02	0.88059D 00
30.	0.25973D-02	0.34858D-02	0.74576D 00
40.	0.26034D-02	0.44834D-02	0.58177D 00
45.	0.26076D-02	0.52803D-02	0.49515D 00
50.	0.26126D-02	0.64124D-02	0.40897D 00
60.	0.26255D-02	0.10661D-01	0.24824D 00
70.	0.26426D-02	0.22763D-01	0.11842D 00
80.	0.26642D-02	0.83901D-01	0.34327D-01
90.	0.26903D-02	0.10218D 01	0.53089D-02
100.	0.27202D-02	0.85295D-01	0.34518D-01
110.	0.27531D-02	0.23679D-01	0.11869D 00
120.	0.27874D-02	0.11315D-01	0.24844D 00
130.	0.28213D-02	0.69261D-02	0.40901D 00
135.	0.28376D-02	0.57478D-02	0.49511D 00
150.	0.28802D-02	0.38668D-02	0.74559D 00
165.	0.29089D-02	0.31236D-02	0.93144D 00
180.	0.29190D-02	0.29190D-02	0.10000D 01

ALPHA = 0.80

THETA	RHO S	RHO P	RHO U
0.	0.25279D-02	0.25279D-02	0.10000D 01
5.	0.25281D-02	0.25484D-02	0.99205D 00
10.	0.25289D-02	0.26115D-02	0.96847D 00
20.	0.25323D-02	0.28845D-02	0.87820D 00
30.	0.25387D-02	0.34274D-02	0.74137D 00
40.	0.25492D-02	0.44341D-02	0.57600D 00
45.	0.25565D-02	0.52404D-02	0.48914D 00
50.	0.256520-02	0.63883D-02	0.40308D 00
60.	0.25880D-02	0.10712D-01	0.24356D 00
70.	0.26187D-02	0.23088D-01	0.11574D 00
80.	0.26581D-02	0.85941D-01	0.33498D-01
90.	0.27061D-02	0.10407D 01	0.52919D-02
100.	0.27619D-02	0.88635D-01	0.33829D-01
110.	0.28239D-02	0.24859D-01	0.11610D 00
120.	0.28894D-02	0.11971D-01	0.24355D 00
130.	0.29549D-02	0.73736D-02	0.40251D 00
135.	0.29865D-02	0.61349D-02	0.48833D 00
150.	0.30700D-02	0.41510D-02	0.74038D 00
165.	0.31267D-02	0.33640D-02	0.92968D 00
180.	0.31468D-02	0.31468D-02	0.10000D 01

ALPHA = 1.00

THETA	RHO S	RHO P	RHO U
0.	0.24465D-02	0.24465D-02	0.10000D 01
5.	0.24468D-02	0.24671D-02	0.99182D 00
10.	0.24480D-02	0.25301D-02	0.96760D 00
20.	0.24528D-02	0.28039D-02	0.87509D 00
30.	0.24622D-02	0.33500D-02	0.73563D 00
40.	0.24780D-02	0.43677D-02	0.56841D 00
45.	0.24889D-02	0.51862D-02	0.48120D 00
50.	0.25023D-02	0.63552D-02	0.39526D 00
60.	0.25377D-02	0.10783D-01	0.23727D 00
70.	0.25863D-02	0.23557D-01	0.11209D 00
80.	0.26496D-02	0.88946D-01	0.32353D-01
90.	0.27282D-02	0.10681D 01	0.52681D-02
100.	0.28214D-02	0.93676D-01	0.32847D-01
110.	0.29266D-02	0.26670D-01	0.11233D 00
120.	0.30395D-02	0.12988D-01	0.23635D 00
130.	0.31543D-02	0.80693D-02	0.39282D 00
135.	0.32101D-02	0.67373D-02	0.47815D 00
150.	0.33598D-02	0.45930D-02	0.73241D 00
165.	0.34629D-02	0.37368D-02	0.92695D 00
180.	0.34997D-02	0.34997D-02	0.10000D 01

ALPHA = 0.10

THETA	RHO S	RHO P	RHO U
0.	0.60476D-02	0.60476D-02	0.10000D 01
5.	0.60476D-02	0.60937D-02	0.99249D 00
10.	0.60476D-02	0.62346D-02	0.97019D 00
20.	0.60477D-02	0.68441D-02	0.88435D 00
30.	0.60479D-02	0.80493D-02	0.75286D 00
40.	0.60482D-02	0.10266D-01	0.59160D 00
45.	0.60485D-02	0.12029D-01	0.50582D 00
50.	0.60487D-02	0.14522D-01	0.42004D 00
60.	0.60494D-02	0.23781D-01	0.25887D 00
70.	0.60504D-02	0.49498D-01	0.12751D 00
80.	0.60515D-02	0.16813D 00	0.41793D-01
90.	0.60530D-02	0.10006D 01	0.12030D-01
100.	0.60546D-02	0.16817D 00	0.41804D-01
110.	0.60564D-02	0.49542D-01	0.12753D 00
120.	0.60583D-02	0.23814D-01	0.25889D 00
130.	0.60601D-02	0.14548D-01	0.42007D 00
135.	0.60610D-02	0.12053D-01	0.50584D 00
150.	0.60632D-02	0.80695D-02	0.75287D 00
165.	0.60647D-02	0.64977D-02	0.93377D 00
180.	0.60653D-02	0.60653D-02	0.10000D 01

ALPHA = 0.20

THETA	RHO S	RHO P	RHO U
0.	0.60344D-02	0.60344D-02	0.10000D 01
5.	0.60344D-02	0.60805D-02	0.99247D 00
10.	0.60345D-02	0.62216D-02	0.97012D 00
20.	0.60349D-02	0.68314D-02	0.88410D 00
30.	0.60357D-02	0.80377D-02	0.75241D 00
40.	0.60369D-02	0.10257D-01	0.59102D 00
45.	0.60378D-02	0.12022D-01	0.50521D 00
50.	0.60389D-02	0.14518D-01	0.41945D 00
60.	0.60417D-02	0.23793D-01	0.25840D 00
70.	0.60454D-02	0.49566D-01	0.12724D 00
80.	0.60502D-02	0.16849D 00	0.41705D-01
90.	0.60559D-02	0.10023D 01	0.12025D-01
100.	0.60625D-02	0.16868D 00	0.41751D-01
110.	0.60697D-02	0.49740D-01	0.12733D 00
120.	0.60771D-02	0.23926D-01	0.25850D 00
130.	0.60845D-02	0.14626D-01	0.41955D 00
135.	0.60880D-02	0.12120D-01	0.50530D 00
150.	0.60971D-02	0.81192D-02	0.75247D 00
165.	0.61033D-02	0.65399D-02	0.93364D 00
180.	0.61054D-02	0.61054D-02	0.10000D 01

ALPHA = 0.40

THETA	RHO S	RHO P	RHO U
0.	0.59816D-02	0.59816D-02	0.10000D 01
5.	0.59817D-02	0.60277D-02	0.99240D 00
10.	0.59820D-02	0.61692D-02	0.96984D 00
20.	0.59835D-02	0.67808D-02	0.88312D 00
30.	0.59865D-02	0.79913D-02	0.75062D 00
40.	0.59915D-02	0.10221D-01	0.58868D 00
45.	0.59950D-02	0.11995D-01	0.50279D 00
50.	0.59992D-02	0.14506D-01	0.41708D 00
60.	0.60103D-02	0.23845D-01	0.25652D 00
70.	0.60254D-02	0.49841D-01	0.12616D 00
80.	0.60445D-02	0.17001D 00	0.41348D-01
90.	0.60678D-02	0.10094D 01	0.12006D-01
100.	0.60946D-02	0.17076D 00	0.41532D-01
110.	0.61240D-02	0.50559D-01	0.12647D 00
120.	0.61547D-02	0.24392D-01	0.25690D 00
130.	0.61850D-02	0.14946D-01	0.41743D 00
135.	0.61995D-02	0.12398D-01	0.50310D 00
150.	0.62374D-02	0.83250D-02	0.75079D 00
165.	0.62628D-02	0.67150D-02	0.93307D 00
180.	0.62718D-02	0.62718D-02	0.10000D 01

ALPHA = 0.60

THETA	RHO S	RHO P	RHO U
0.	0.58932D-02	0.58932D-02	0.10000D 01
5.	0.58934D-02	0.59395D-02	0.99228D 00
10.	0.58942D-02	0.60815D-02	0.96938D 00
20.	0.58974D-02	0.66957D-02	0.88148D 00
30.	0.59039D-02	0.79127D-02	0.74761D 00
40.	0.59149D-02	0.10158D-01	0.58474D 00
45.	0.59226D-02	0.11948D-01	0.49869D 00
50.	0.59321D-02	0.14483D-01	0.41306D 00
60.	0.59570D-02	0.23936D-01	0.25332D 00
70.	0.59910D-02	0.50324D-01	0.12430D 00
80.	0.60348D-02	0.17270D 00	0.40734D-01
90.	0.60884D-02	0.10218D 01	0.11974D-01
100.	0.61506D-02	0.17447D 00	0.41150D-01
110.	0.62194D-02	0.52028D-01	0.12498D 00
120.	0.62916D-02	0.25231D-01	0.25406D 00
130.	0.63635D-02	0.15523D-01	0.41366D 00
135.	0.63979D-02	0.12900D-01	0.49917D 00
150.	0.64886D-02	0.86961D-02	0.74779D 00
165.	0.65497D-02	0.70305D-02	0.93206D 00
180.	0.65713D-02	0.65713D-02	0.10000D 01

ALPHA = 0.80

THETA	RHO S	RHO P	RHO U
0.	0.57688D-02	0.57688D-02	0.10000D 01
5.	0.57692D-02	0.58154D-02	0.99211D 00
10.	0.57705D-02	0.59579D-02	0.96873D 00
20.	0.57759D-02	0.65750D-02	0.87916D 00
30.	0.57868D-02	0.78004D-02	0.74335D 00
40.	0.58058D-02	0.10068D-01	0.57913D 00
45.	0.58192D-02	0.11879D-01	0.49283D 00
50.	0.58357D-02	0.14451D-01	0.40730D 00
60.	0.58799D-02	0.24070D-01	0.24870D 00
70.	0.59410D-02	0.51056D-01	0.12158D 00
80.	0.60206D-02	0.17682D 00	0.39829D-01
90.	0.61190D-02	0.10407D 01	0.11926D-01
100.	0.62348D-02	0.18022D 00	0.40576D-01
110.	0.63641D-02	0.54334D-01	0.12271D 00
120.	0.65015D-02	0.26555D-01	0.24971D 00
130.	0.66394D-02	0.16438D-01	0.40783D 00
135.	0.67059D-02	0.13694D-01	0.49309D 00
150.	0.68824D-02	0.92841D-02	0.74308D 00
165.	0.70024D-02	0.75296D-02	0.93047D 00
180.	0.70450D-02	0.70450D-02	0.10000D 01

ALPHA = 1.00

THETA	RHO S	RHO P	RHO U
0.	0.56080D-02	0.56080D-02	0.10000D 01
5.	0.56085D-02	0.56546D-02	0.99190D 00
10.	0.56103D-02	0.57975D-02	0.96788D 00
20.	0.56180D-02	0.64173D-02	0.87614D 00
30.	0.56339D-02	0.76517D-02	0.73778D 00
40.	0.56621D-02	0.99456D-02	0.57173D 00
45.	0.56825D-02	0.11785D-01	0.48508D 00
50.	0.57078D-02	0.14406D-01	0.39965D 00
60.	0.57764D-02	0.24260D-01	0.24248D 00
70.	0.58729D-02	0.52107D-01	0.11789D 00
80.	0.60009D-02	0.18288D 00	0.38582D-01
90.	0.61619D-02	0.10679D 01	0.11859D-01
100.	0.63542D-02	0.18877D 00	0.39762D-01
110.	0.65725D-02	0.57825D-01	0.11945D 00
120.	0.68079D-02	0.28579D-01	0.24337D 00
130.	0.70477D-02	0.17843D-01	0.39922D 00
135.	0.71645D-02	0.14916D-01	0.48403D 00
150.	0.74777D-02	0.10188D-01	0.73597D 00
165.	0.76934D-02	0.82950D-02	0.92803D 00
180.	0.77705D-02	0.77705D-02	0.10000D 01

ALPHA = 0.10

THETA	RHO S	RHO P	RHO U
0.	0.10734D-01	0.10734D-01	0.10000D 01
5.	0.10734D-01	0.10815D-01	0.99256D 00
10.	0.10734D-01	0.11064D-01	0.97046D 00
20.	0.10734D-01	0.12140D-01	0.88543D 00
30.	0.10734D-01	0.14264D-01	0.75517D 00
40.	0.10734D-01	0.18161D-01	0.59542D 00
45.	0.10735D-01	0.21250D-01	0.51043D 00
50.	0.10735D-01	0.25603D-01	0.42545D 00
60.	0.10736D-01	0.41629D-01	0.26578D 00
70.	0.10738D-01	0.84969D-01	0.13565D 00
80.	0.10740D-01	0.26493D 00	0.50733D-01
90.	0.10742D-01	0.10006D 01	0.21249D-01
100.	0.10745D-01	0.26496D 00	0.50751D-01
110.	0.10747D-01	0.85030D-01	0.13569D 00
120.	0.10751D-01	0.41680D-01	0.26583D 00
130.	0.10754D-01	0.25646D-01	0.42550D 00
135.	0.10755D-01	0.21289D-01	0.51047D 00
150.	0.10759D-01	0.14296D-01	0.75519D 00
165.	0.10761D-01	0.11526D-01	0.93440D 00
180.	0.10762D-01	0.10762D-01	0.10000D 01

ALPHA = 0.20

THETA	RHO S	RHO P	RHO U
0.	0.10713D-01	0.10713D-01	0.10000D 01
5.	0.10713D-01	0.10795D-01	0.99254D 00
10.	0.10713D-01	0.11044D-01	0.97040D 00
20.	0.10714D-01	0.12120D-01	0.88519D 00
30.	0.10715D-01	0.14246D-01	0.75474D 00
40.	0.10717D-01	0.18148D-01	0.59485D 00
45.	0.10718D-01	0.21241D-01	0.50984D 00
50.	0.10719D-01	0.25601D-01	0.42488D 00
60.	0.10724D-01	0.41654D-01	0.26532D 00
70.	0.10729D-01	0.85087D-01	0.13538D 00
80.	0.10737D-01	0.26550D 00	0.50634D-01
90.	0.10746D-01	0.10023D 01	0.21240D-01
100.	0.10757D-01	0.26563D 00	0.50707D-01
110.	0.10769D-01	0.85333D-01	0.13551D 00
120.	0.10781D-01	0.41859D-01	0.26548D 00
130.	0.10794D-01	0.25771D-01	0.42504D 00
135.	0.10799D-01	0.21398D-01	0.50999D 00
150.	0.10815D-01	0.14378D-01	0.75483D 00
165.	0.10825D-01	0.11596D-01	0.93427D 00
180.	0.10829D-01	0.10829D-01	0.10000D 01

ALPHA = 0.40

THETA	RHO S	RHO P	RHO U
0.	0.10632D-01	0.10632D-01	0.10000D 01
5.	0.10632D-01	0.10713D-01	0.99247D 00
10.	0.10632D-01	0.10963D-01	0.97013D 00
20.	0.10634D-01	0.12043D-01	0.88424D 00
30.	0.10638D-01	0.14176D-01	0.75300D 00
40.	0.10644D-01	0.18096D-01	0.59257D 00
45.	0.10649D-01	0.21205D-01	0.50746D 00
50.	0.10656D-01	0.25591D-01	0.42254D 00
60.	0.10672D-01	0.41757D-01	0.26344D 00
70.	0.10696D-01	0.85568D-01	0.13426D 00
80.	0.10726D-01	0.26786D 00	0.50232D-01
90.	0.10764D-01	0.10094D 01	0.21200D-01
100.	0.10808D-01	0.26840D 00	0.50528D-01
110.	0.10856D-01	0.86583D-01	0.13478D 00
120.	0.10907D-01	0.42599D-01	0.26407D 00
130.	0.10958D-01	0.26288D-01	0.42315D 00
135.	0.10982D-01	0.21849D-01	0.50802D 00
150.	0.11046D-01	0.14716D-01	0.75332D 00
165.	0.11088D-01	0.11884D-01	0.93376D 00
180.	0.11104D-01	0.11104D-01	0.10000D 01

ALPHA = 0.60

THETA	RHO S	RHO P	RHO U
0.	0.10495D-01	0.10495D-01	0.10000D 01
5.	0.10495D-01	0.10577D-01	0.99236D 00
10.	0.10496D-01	0.10828D-01	0.96968D 00
20.	0.10500D-01	0.11912D-01	0.88265D 00
30.	0.10508D-01	0.14058D-01	0.75007D 00
40.	0.10523D-01	0.18007D-01	0.58872D 00
45.	0.10534D-01	0.21143D-01	0.50345D 00
50.	0.10548D-01	0.25573D-01	0.41859D 00
60.	0.10585D-01	0.41936D-01	0.26025D 00
70.	0.10638D-01	0.86411D-01	0.13234D 00
80.	0.10708D-01	0.27203D 00	0.49541D-01
90.	0.10794D-01	0.10218D 01	0.21131D-01
100.	0.10896D-01	0.27330D 00	0.50216D-01
110.	0.11009D-01	0.88811D-01	0.13350D 00
120.	0.11129D-01	0.43923D-01	0.26159D 00
130.	0.11248D-01	0.27217D-01	0.41981D 00
135.	0.11306D-01	0.22660D-01	0.50453D 00
150.	0.11457D-01	0.15322D-01	0.75063D 00
165.	0.11560D-01	0.12402D-01	0.93285D 00
180.	0.11596D-01	0.11596D-01	0.10000D 01

ALPHA = 0.80

THETA	RHO S	RHO P	RHO U
0.	0.10302D-01	0.10302D-01	0.10000D 01
5.	0.10303D-01	0.10385D-01	0.99220D 00
10.	0.10304D-01	0.10637D-01	0.96905D 00
20.	0.10310D-01	0.11727D-01	0.88040D 00
30.	0.10324D-01	0.13889D-01	0.74593D 00
40.	0.10350D-01	0.17877D-01	0.58324D 00
45.	0.10369D-01	0.21053D-01	0.49771D 00
50.	0.10393D-01	0.25547D-01	0.41292D 00
60.	0.10460D-01	0.42201D-01	0.25564D 00
70.	0.10554D-01	0.87686D-01	0.12955D 00
80.	0.10681D-01	0.27842D 00	0.48525D-01
90.	0.10840D-01	0.10406D 01	0.21028D-01
100.	0.11028D-01	0.28083D 00	0.49749D-01
110.	0.11240D-01	0.92277D-01	0.13157D 00
120.	0.11467D-01	0.45997D-01	0.25780D 00
130.	0.11695D-01	0.28676D-01	0.41468D 00
135.	0.11806D-01	0.23935D-01	0.49915D 00
150.	0.12099D-01	0.16276D-01	0.74644D 00
165.	0.12299D-01	0.13216D-01	0.93143D 00
180.	0.12370D-01	0.12370D-01	0.10000D 01

ALPHA = 1.00

THETA	RHO S	RHO P	RHO U
0.	0.10053D-01	0.10053D-01	0.10000D 01
5.	0.10053D-01	0.10136D-01	0.99198D 00
10.	0.10055D-01	0.10389D-01	0.96823D 00
20.	0.10064D-01	0.11485D-01	0.87747D 00
30.	0.10083D-01	0.13665D-01	0.74050D 00
40.	0.10122D-01	0.17704D-01	0.57601D 00
45.	0.10150D-01	0.20931D-01	0.49012D 00
50.	0.10187D-01	0.25511D-01	0.40539D 00
60.	0.10291D-01	0.42574D-01	0.24944D 00
70.	0.10441D-01	0.89513D-01	0.12577D 00
80.	0.10644D-01	0.28777D 00	0.47130D-01
90.	0.10903D-01	0.10678D 01	0.20886D-01
100.	0.11215D-01	0.29185D 00	0.49090D-01
110.	0.11571D-01	0.97457D-01	0.12881D 00
120.	0.11957D-01	0.49129D-01	0.25232D 00
130.	0.12351D-01	0.30890D-01	0.40716D 00
135.	0.12543D-01	0.25871D-01	0.49122D 00
150.	0.13059D-01	0.17724D-01	0.74019D 00
165.	0.13415D-01	0.14450D-01	0.92928D 00
180.	0.13542D-01	0.13542D-01	0.10000D 01

ALPHA = 0.10

THETA	RHO S	RHO P	RHO U
0.	0.14881D-03	0.14881D-03	0.10000D 01
5.	0.14881D-03	0.14995D-03	0.99240D 00
10.	0.14881D-03	0.15344D-03	0.96983D 00
20.	0.14882D-03	0.16854D-03	0.88297D 00
30.	0.14883D-03	0.19847D-03	0.74991D 00
40.	0.14884D-03	0.25370D-03	0.58674D 00
45.	0.14885D-03	0.29778D-03	0.49993D 00
50.	0.14886D-03	0.36039D-03	0.41314D 00
60.	0.14889D-03	0.59567D-03	0.25006D 00
70.	0.14892D-03	0.12726D-02	0.11715D 00
80.	0.14896D-03	0.49206D-02	0.30418D-01
90.	0.14901D-03	0.10006D 01	0.29789D-03
100.	0.14906D-03	0.49239D-02	0.30417D-01
110.	0.14912D-03	0.12743D-02	0.11715D 00
120.	0.14917D-03	0.59681D-03	0.25006D 00
130.	0.14923D-03	0.36128D-03	0.41314D 00
135.	0.14925D-03	0.29859D-03	0.49993D 00
150.	0.14932D-03	0.19912D-03	0.74991D 00
165.	0.14936D-03	0.16009D-03	0.93298D 00
180.	0.14938D-03	0.14938D-03	0.10000D 01

ALPHA = 0.20

THETA	RHO S	RHO P	RHO U
0.	0.14835D-03	0.14835D-03	0.10000D 01
5.	0.14836D-03	0.14949D-03	0.99238D 00
10.	0.14836D-03	0.15299D-03	0.96975D 00
20.	0.14838D-03	0.16810D-03	0.88270D 00
30.	0.14842D-03	0.19805D-03	0.74943D 00
40.	0.14848D-03	0.25335D-03	0.58611D 00
45.	0.14852D-03	0.29750D-03	0.49929D 00
50.	0.14856D-03	0.36021D-03	0.41252D 00
60.	0.14867D-03	0.59594D-03	0.24958D 00
70.	0.14880D-03	0.12745D-02	0.11689D 00
80.	0.14897D-03	0.49329D-02	0.30343D-01
90.	0.14915D-03	0.10023D 01	0.29792D-03
100.	0.14936D-03	0.49465D-02	0.30340D-01
110.	0.14958D-03	0.12813D-02	0.11688D 00
120.	0.14981D-03	0.60055D-03	0.24957D 00
130.	0.15003D-03	0.36378D-03	0.41251D 00
135.	0.15013D-03	0.30075D-03	0.49928D 00
150.	0.15040D-03	0.20070D-03	0.74942D 00
165.	0.15058D-03	0.16143D-03	0.93281D 00
180.	0.15064D-03	0.15064D-03	0.10000D 01

ALPHA = 0.40

THETA	RHO S	RHO P	RHO U
0.	0.14653D-03	0.14653D-03	0.10000D 01
5.	0.14654D-03	0.14767D-03	0.99230D 00
10.	0.14656D-03	0.15118D-03	0.96945D 00
20.	0.14664D-03	0.16633D-03	0.88164D 00
30.	0.14680D-03	0.19639D-03	0.74749D 00
40.	0.14703D-03	0.25196D-03	0.58360D 00
45.	0.14718D-03	0.29636D-03	0.49669D 00
50.	0.14735D-03	0.35947D-03	0.41000D 00
60.	0.14778D-03	0.59705D-03	0.24762D 00
70.	0.14832D-03	0.12822D-02	0.11581D 00
80.	0.14898D-03	0.49838D-02	0.30037D-01
90.	0.14974D-03	0.10093D 01	0.29806D-03
100.	0.15060D-03	0.50401D-02	0.30026D-01
110.	0.15151D-03	0.13102D-02	0.11577D 00
120.	0.15245D-03	0.61611D-03	0.24755D 00
130.	0.15336D-03	0.37423D-03	0.40988D 00
135.	0.15379D-03	0.30975D-03	0.49657D 00
150.	0.15491D-03	0.20728D-03	0.74738D 00
165.	0.15565D-03	0.16699D-03	0.93213D 00
180.	0.15591D-03	0.15591D-03	0.10000D 01

ALPHA = 0.60

THETA	RHO S	RHO P	RHO U
0.	0.14349D-03	0.14349D-03	0.10000D 01
5.	0.14351D-03	0.14464D-03	0.99217D 00
10.	0.14355D-03	0.14815D-03	0.96895D 00
20.	0.14374D-03	0.16337D-03	0.87985D 00
30.	0.14407D-03	0.19359D-03	0.74424D 00
40.	0.14458D-03	0.24958D-03	0.57937D 00
45.	0.14491D-03	0.29439D-03	0.49231D 00
50.	0.14530D-03	0.35820D-03	0.40573D 00
60.	0.14626D-03	0.59899D-03	0.24429D 00
70.	0.14749D-03	0.12958D-02	0.11395D 00
80.	0.14900D-03	0.50741D-02	0.29510D-01
90.	0.15077D-03	0.10216D 01	0.29831D-03
100.	0.15276D-03	0.52092D-02	0.29474D-01
110.	0.15492D-03	0.13629D-02	0.11381D 00
120.	0.15714D-03	0.64446D-03	0.24395D 00
130.	0.15932D-03	0.39330D-03	0.40518D 00
135.	0.16036D-03	0.32618D-03	0.49170D 00
150.	0.16307D-03	0.21928D-03	0.74369D 00
165.	0.16488D-03	0.17712D-03	0.93089D 00
180.	0.16552D-03	0.16552D-03	0.10000D 01

ALPHA = 0.80

THETA	RHO S	RHO P	RHO U
0.	0.13923D-03	0.13923D-03	0.10000D 01
5.	0.13926D-03	0.14038D-03	0.99199D 00
10.	0.13934D-03	0.14391D-03	0.96825D 00
20.	0.13965D-03	0.15918D-03	0.87735D 00
30.	0.14022D-03	0.18959D-03	0.73965D 00
40.	0.14110D-03	0.24613D-03	0.57335D 00
45.	0.14168D-03	0.29153D-03	0.48606D 00
50.	0.14236D-03	0.35633D-03	0.39962D 00
60.	0.14407D-03	0.60189D-03	0.23948D 00
70.	0.14629D-03	0.13165D-02	0.11125D 00
80.	0.14903D-03	0.52138D-02	0.28729D-01
90.	0.15231D-03	0.10403D 01	0.29867D-03
100.	0.15605D-03	0.54781D-02	0.28637D-01
110.	0.16014D-03	0.14473D-02	0.11079D 00
120.	0.16443D-03	0.69020D-03	0.23835D 00
130.	0.16868D-03	0.42414D-03	0.39780D 00
135.	0.17072D-03	0.35278D-03	0.48402D 00
150.	0.17610D-03	0.23870D-03	0.73779D 00
165.	0.17974D-03	0.19350D-03	0.92890D 00
180.	0.18103D-03	0.18103D-03	0.10000D 01

ALPHA = 1.00

THETA	RHO S	RHO P	RHO U
0.	0.13376D-03	0.13376D-03	0.10000D 01
5.	0.13380D-03	0.13491D-03	0.99176D 00
10.	0.13391D-03	0.13843D-03	0.96734D 00
20.	0.13437D-03	0.15373D-03	0.87411D 00
30.	0.13522D-03	0.18431D-03	0.73368D 00
40.	0.13654D-03	0.24149D-03	0.56546D 00
45.	0.13742D-03	0.28765D-03	0.47782D 00
50.	0.13847D-03	0.35376D-03	0.39150D 00
60.	0.14113D-03	0.60598D-03	0.23300D 00
70.	0.14464D-03	0.13464D-02	0.10755D 00
80.	0.14908D-03	0.54208D-02	0.27646D-01
90.	0.15448D-03	0.10671D 01	0.29919D-03
100.	0.16077D-03	0.58937D-02	0.27434D-01
110.	0.16779D-03	0.15796D-02	0.10638D 00
120.	0.17528D-03	0.76247D-03	0.23002D 00
130.	0.18286D-03	0.47310D-03	0.38662D 00
135.	0.18654D-03	0.39504D-03	0.47229D 00
150.	0.19638D-03	0.26954D-03	0.72861D 00
165.	0.20314D-03	0.21944D-03	0.92575D 00
180.	0.20556D-03	0.20556D-03	0.10000D 01

ALPHA = 0.10

THETA	RHO S	RHO P	RHO U
0.	0.60163D-03	0.60163D-03	0.10000D 01
5.	0.60163D-03	0.60624D-03	0.99241D 00
10.	0.60164D-03	0.62035D-03	0.96986D 00
20.	0.60166D-03	0.68137D-03	0.88308D 00
30.	0.60169D-03	0.80226D-03	0.75014D 00
40.	0.60175D-03	0.10254D-02	0.58712D 00
45.	0.60178D-03	0.12033D-02	0.50039D 00
50.	0.60182D-03	0.14560D-02	0.41368D 00
60.	0.60192D-03	0.24049D-02	0.25075D 00
70.	0.60205D-03	0.51272D-02	0.11795D 00
80.	0.60221D-03	0.19607D-01	0.31297D-01
90.	0.60239D-03	0.10006D 01	0.12037D-02
100.	0.60259D-03	0.19620D-01	0.31296D-01
110.	0.60280D-03	0.51336D-02	0.11795D 00
120.	0.60302D-03	0.24093D-02	0.25074D 00
130.	0.60323D-03	0.14595D-02	0.41368D 00
135.	0.60333D-03	0.12064D-02	0.50039D 00
150.	0.60359D-03	0.80480D-03	0.75014D 00
165.	0.60377D-03	0.64712D-03	0.93304D 00
180.	0.60383D-03	0.60383D-03	0.10000D 01

ALPHA = 0.20

THETA	RHO S	RHO P	RHO U
0.	0.59995D-03	0.59995D-03	0.10000D 01
5.	0.59996D-03	0.60456D-03	0.99239D 00
10.	0.59998D-03	0.61868D-03	0.96978D 00
20.	0.60006D-03	0.67976D-03	0.88282D 00
30.	0.60020D-03	0.80077D-03	0.74967D 00
40.	0.60041D-03	0.10241D-02	0.58651D 00
45.	0.60055D-03	0.12024D-02	0.49976D 00
50.	0.60071D-03	0.14555D-02	0.41307D 00
60.	0.60112D-03	0.24062D-02	0.25027D 00
70.	0.60163D-03	0.51350D-02	0.11769D 00
80.	0.60225D-03	0.19656D-01	0.31223D-01
90.	0.60297D-03	0.10023D 01	0.12038D-02
100.	0.60378D-03	0.19707D-01	0.31222D-01
110.	0.60464D-03	0.51609D-02	0.11769D 00
120.	0.60553D-03	0.24239D-02	0.25027D 00
130.	0.60638D-03	0.14693D-02	0.41306D 00
135.	0.60679D-03	0.12149D-02	0.49975D 00
150.	0.60784D-03	0.81098D-03	0.74966D 00
165.	0.60854D-03	0.65235D-03	0.93288D 00
180.	0.60878D-03	0.60878D-03	0.10000D 01

ALPHA = 0.40

THETA	RHO S	RHO P	RHO U
0.	0.59323D-03	0.59323D-03	0.10000D 01
5.	0.59325D-03	0.59785D-03	0.99231D 00
10.	0.59333D-03	0.61201D-03	0.96949D 00
20.	0.59364D-03	0.67328D-03	0.88177D 00
30.	0.59419D-03	0.79477D-03	0.74778D 00
40.	0.59504D-03	0.10192D-02	0.58405D 00
45.	0.59559D-03	0.11986D-02	0.49722D 00
50.	0.59624D-03	0.14534D-02	0.41060D 00
60.	0.59786D-03	0.24116D-02	0.24836D 00
70.	0.59991D-03	0.51669D-02	0.11664D 00
80.	0.60242D-03	0.19856D-01	0.30923D-01
90.	0.60536D-03	0.10093D 01	0.12044D-02
100.	0.60866D-03	0.20069D-01	0.30918D-01
110.	0.61221D-03	0.52744D-02	0.11661D 00
120.	0.61584D-03	0.24849D-02	0.24830D 00
130.	0.61940D-03	0.15102D-02	0.41051D 00
135.	0.62108D-03	0.12501D-02	0.49712D 00
150.	0.62545D-03	0.83670D-03	0.74768D 00
165.	0.62837D-03	0.67409D-03	0.93222D 00
180.	0.62939D-03	0.62939D-03	0.10000D 01

ALPHA = 0.60

THETA	RHO S	RHO P	RHO U
0.	0.58202D-03	0.58202D-03	0.10000D 01
5.	0.58207D-03	0.58666D-03	0.99219D 00
10.	0.58223D-03	0.60087D-03	0.96900D 00
20.	0.58290D-03	0.66242D-03	0.88003D 00
30.	0.58412D-03	0.78463D-03	0.74460D 00
40.	0.58600D-03	0.10109D-02	0.57991D 00
45.	0.58723D-03	0.11920D-02	0.49294D 00
50.	0.58868D-03	0.14497D-02	0.40643D 00
60.	0.59230D-03	0.24209D-02	0.24511D 00
70.	0.59697D-03	0.52230D-02	0.11483D 00
80.	0.60273D-03	0.20211D-01	0.30406D-01
90.	0.60952D-03	0.10216D 01	0.12054D-02
100.	0.61722D-03	0.20722D-01	0.30385D-01
110.	0.62556D-03	0.54800D-02	0.11471D 00
120.	0.63419D-03	0.25957D-02	0.24480D 00
130.	0.64268D-03	0.15847D-02	0.40594D 00
135.	0.64672D-03	0.13143D-02	0.49239D 00
150.	0.65729D-03	0.88354D-03	0.74409D 00
165.	0.66437D-03	0.71364D-03	0.93101D 00
180.	0.66686D-03	0.66686D-03	0.10000D 01

ALPHA = 0.80

THETA	RHO S	RHO P	RHO U
0.	0.56631D-03	0.56631D-03	0.10000D 01
5.	0.56640D-03	0.57097D-03	0.99201D 00
10.	0.56667D-03	0.58523D-03	0.96831D 00
20.	0.56782D-03	0.64708D-03	0.87758D 00
30.	0.56990D-03	0.77017D-03	0.74011D 00
40.	0.57315D-03	0.99888D-03	0.57403D 00
45.	0.57530D-03	0.11824D-02	0.48684D 00
50.	0.57785D-03	0.14442D-02	0.40046D 00
60.	0.58428D-03	0.24349D-02	0.24040D 00
70.	0.59268D-03	0.53082D-02	0.11218D 00
80.	0.60317D-03	0.20758D-01	0.29642D-01
90.	0.61574D-03	0.10402D 01	0.12070D-02
100.	0.63016D-03	0.21755D-01	0.29578D-01
110.	0.64599D-03	0.58085D-02	0.11179D 00
120.	0.66259D-03	0.27736D-02	0.23939D 00
130.	0.67910D-03	0.17046D-02	0.39880D 00
135.	0.68702D-03	0.14177D-02	0.48495D 00
150.	0.70794D-03	0.95900D-03	0.73839D 00
165.	0.72209D-03	0.77725D-03	0.92908D 00
180.	0.72710D-03	0.72710D-03	0.10000D 01

ALPHA = 1.00

THETA	RHO S	RHO P	RHO U
0.	0.54610D-03	0.54610D-03	0.10000D 01
5.	0.54623D-03	0.55076D-03	0.99178D 00
10.	0.54663D-03	0.56505D-03	0.96742D 00
20.	0.54831D-03	0.62711D-03	0.87442D 00
30.	0.55139D-03	0.75109D-03	0.73427D 00
40.	0.55628D-03	0.98269D-03	0.56632D 00
45.	0.55956D-03	0.11694D-02	0.47878D 00
50.	0.56348D-03	0.14367D-02	0.39254D 00
60.	0.57350D-03	0.24546D-02	0.23408D 00
70.	0.58682D-03	0.54310D-02	0.10857D 00
80.	0.60379D-03	0.21566D-01	0.28584D-01
90.	0.62449D-03	0.10668D 01	0.12091D-02
100.	0.64870D-03	0.23341D-01	0.28423D-01
110.	0.67580D-03	0.63197D-02	0.10754D 00
120.	0.70472D-03	0.30531D-02	0.23136D 00
130.	0.73399D-03	0.18937D-02	0.38803D 00
135.	0.74820D-03	0.15809D-02	0.47367D 00
150.	0.78624D-03	0.10780D-02	0.72956D 00
165.	0.81240D-03	0.87732D-03	0.92606D 00
180.	0.82174D-03	0.82174D-03	0.10000D 01

ALPHA = 0.10

THETA	RHO S	RHO P	RHO U
0.	0.24400D-02	0.24400D-02	0.10000D 01
5.	0.24400D-02	0.24587D-02	0.99243D 00
10.	0.24400D-02	0.25158D-02	0.96997D 00
20.	0.24401D-02	0.27627D-02	0.88351D 00
30.	0.24402D-02	0.32516D-02	0.75107D 00
40.	0.24404D-02	0.41529D-02	0.58864D 00
45.	0.24405D-02	0.48711D-02	0.50223D 00
50.	0.24406D-02	0.58894D-02	0.41584D 00
60.	0.24410D-02	0.96989D-02	0.25350D 00
70.	0.24415D-02	0.20508D-01	0.12120D 00
80.	0.24420D-02	0.75145D-01	0.34854D-01
90.	0.24427D-02	0.10006D 01	0.48721D-02
100.	0.24434D-02	0.75186D-01	0.34857D-01
110.	0.24442D-02	0.20531D-01	0.12120D 00
120.	0.24451D-02	0.97149D-02	0.25351D 00
130.	0.24459D-02	0.59020D-02	0.41584D 00
135.	0.24463D-02	0.48825D-02	0.50224D 00
150.	0.24473D-02	0.32610D-02	0.75107D 00
165.	0.24479D-02	0.26233D-02	0.93329D 00
180.	0.24481D-02	0.24481D-02	0.10000D 01

ALPHA = 0.20

THETA	RHO S	RHO P	RHO U
0.	0.24343D-02	0.24343D-02	0.10000D 01
5.	0.24344D-02	0.24530D-02	0.99242D 00
10.	0.24344D-02	0.25102D-02	0.96990D 00
20.	0.24347D-02	0.27574D-02	0.88326D 00
30.	0.24351D-02	0.32468D-02	0.75061D 00
40.	0.24359D-02	0.41493D-02	0.58805D 00
45.	0.24363D-02	0.48686D-02	0.50163D 00
50.	0.24369D-02	0.58887D-02	0.41525D 00
60.	0.24383D-02	0.97058D-02	0.25304D 00
70.	0.24402D-02	0.20541D-01	0.12094D 00
80.	0.24424D-02	0.75329D-01	0.34781D-01
90.	0.24451D-02	0.10023D 01	0.48727D-02
100.	0.24481D-02	0.75493D-01	0.34791D-01
110.	0.24513D-02	0.20633D-01	0.12096D 00
120.	0.24547D-02	0.97706D-02	0.25306D 00
130.	0.24579D-02	0.59394D-02	0.41527D 00
135.	0.24595D-02	0.49148D-02	0.50164D 00
150.	0.24635D-02	0.32846D-02	0.75062D 00
165.	0.24661D-02	0.26433D-02	0.93314D 00
180.	0.24671D-02	0.24671D-02	0.10000D 01

ALPHA = 0.40

THETA	RHO S	RHO P	RHO U
0.	0.24117D-02	0.24117D-02	0.10000D 01
5.	0.24118D-02	0.24304D-02	0.99234D 00
10.	0.24120D-02	0.24878D-02	0.96962D 00
20.	0.24130D-02	0.27359D-02	0.88226D 00
30.	0.24148D-02	0.32275D-02	0.74880D 00
40.	0.24177D-02	0.41349D-02	0.58569D 00
45.	0.24195D-02	0.48588D-02	0.49919D 00
50.	0.24218D-02	0.58859D-02	0.41288D 00
60.	0.24275D-02	0.97341D-02	0.25120D 00
70.	0.24348D-02	0.20675D-01	0.11991D 00
80.	0.24440D-02	0.76084D-01	0.34482D-01
90.	0.24548D-02	0.10093D 01	0.48751D-02
100.	0.24671D-02	0.76763D-01	0.34521D-01
110.	0.24804D-02	0.21059D-01	0.11997D 00
120.	0.24941D-02	0.10002D-01	0.25124D 00
130.	0.25076D-02	0.60949D-02	0.41289D 00
135.	0.25140D-02	0.50489D-02	0.49919D 00
150.	0.25306D-02	0.33826D-02	0.74877D 00
165.	0.25418D-02	0.27262D-02	0.93252D 00
180.	0.25457D-02	0.25457D-02	0.10000D 01

ALPHA = 0.60

THETA	RHO S	RHO P	RHO U
0.	0.23739D-02	0.23739D-02	0.10000D 01
5.	0.23741D-02	0.23927D-02	0.99222D 00
10.	0.23746D-02	0.24504D-02	0.96915D 00
20.	0.23767D-02	0.26999D-02	0.88059D 00
30.	0.23807D-02	0.31950D-02	0.74575D 00
40.	0.23870D-02	0.41104D-02	0.58172D 00
45.	0.23912D-02	0.48419D-02	0.49507D 00
50.	0.23963D-02	0.58811D-02	0.40886D 00
60.	0.24090D-02	0.97831D-02	0.24805D 00
70.	0.24258D-02	0.20910D-01	0.11815D 00
80.	0.24467D-02	0.77418D-01	0.33967D-01
90.	0.24717D-02	0.10215D 01	0.48794D-02
100.	0.25003D-02	0.79037D-01	0.34050D-01
110.	0.25315D-02	0.21825D-01	0.11823D 00
120.	0.25640D-02	0.10419D-01	0.24801D 00
130.	0.25960D-02	0.63766D-02	0.40866D 00
135.	0.26113D-02	0.52917D-02	0.49479D 00
150.	0.26515D-02	0.35602D-02	0.74543D 00
165.	0.26784D-02	0.28762D-02	0.93140D 00
180.	0.26879D-02	0.26879D-02	0.10000D 01

ALPHA = 0.80

THETA	RHO S	RHO P	RHO U
0.	0.23210D-02	0.23210D-02	0.10000D 01
5.	0.23213D-02	0.23399D-02	0.99205D 00
10.	0.23221D-02	0.23979D-02	0.96848D 00
20.	0.23258D-02	0.26491D-02	0.87824D 00
30.	0.23326D-02	0.31486D-02	0.74144D 00
40.	0.23435D-02	0.40750D-02	0.57608D 00
45.	0.23509D-02	0.48173D-02	0.48921D 00
50.	0.23597D-02	0.58741D-02	0.40312D 00
60.	0.23824D-02	0.98562D-02	0.24352D 00
70.	0.24125D-02	0.21266D-01	0.11558D 00
80.	0.24507D-02	0.79465D-01	0.33209D-01
90.	0.24969D-02	0.10400D 01	0.48857D-02
100.	0.25504D-02	0.82602D-01	0.33340D-01
110.	0.26094D-02	0.23038D-01	0.11557D 00
120.	0.26715D-02	0.11084D-01	0.24304D 00
130.	0.27336D-02	0.68264D-02	0.40208D 00
135.	0.27634D-02	0.56798D-02	0.48794D 00
150.	0.28422D-02	0.38438D-02	0.74016D 00
165.	0.28956D-02	0.31156D-02	0.92962D 00
180.	0.29146D-02	0.29146D-02	0.10000D 01

ALPHA = 1.00

THETA	RHO S	RHO P	RHO U
0.	0.22528D-02	0.22528D-02	0.10000D 01
5.	0.22533D-02	0.22718D-02	0.99183D 00
10.	0.22545D-02	0.23301D-02	0.96763D 00
20.	0.22599D-02	0.25829D-02	0.87520D 00
30.	0.22700D-02	0.30874D-02	0.73584D 00
40.	0.22864D-02	0.40276D-02	0.56868D 00
45.	0.22977D-02	0.47840D-02	0.48148D 00
50.	0.23113D-02	0.58644D-02	0.39552D 00
60.	0.23467D-02	0.99586D-02	0.23743D 00
70.	0.23945D-02	0.21776D-01	0.11209D 00
80.	0.24562D-02	0.82460D-01	0.32164D-01
90.	0.25322D-02	0.10663D 01	0.48945D-02
100.	0.26216D-02	0.87992D-01	0.32330D-01
110.	0.27221D-02	0.24901D-01	0.11173D 00
120.	0.28297D-02	0.12114D-01	0.23574D 00
130.	0.29387D-02	0.75260D-02	0.39226D 00
135.	0.29917D-02	0.62841D-02	0.47765D 00
150.	0.31336D-02	0.42851D-02	0.73212D 00
165.	0.32312D-02	0.34870D-02	0.92686D 00
180.	0.32660D-02	0.32660D-02	0.10000D 01

ALPHA = 0.10

THETA	RHO S	RHO P	RHO U
0.	0.55149D-02	0.55149D-02	0.10000D 01
5.	0.55149D-02	0.55570D-02	0.99248D 00
10.	0.55150D-02	0.56856D-02	0.97015D 00
20.	0.55151D-02	0.62417D-02	0.88423D 00
30.	0.55153D-02	0.73416D-02	0.75260D 00
40.	0.55156D-02	0.93657D-02	0.59117D 00
45.	0.55158D-02	0.10975D-01	0.50529D 00
50.	0.55161D-02	0.13253D-01	0.41943D 00
60.	0.55168D-02	0.21721D-01	0.25808D 00
70.	0.55177D-02	0.45314D-01	0.12659D 00
80.	0.55189D-02	0.15555D 00	0.40774D-01
90.	0.55203D-02	0.10006D 01	0.10977D-01
100.	0.55218D-02	0.15561D 00	0.40781D-01
110.	0.55235D-02	0.45358D-01	0.12660D 00
120.	0.55253D-02	0.21753D-01	0.25810D 00
130.	0.55270D-02	0.13279D-01	0.41944D 00
135.	0.55278D-02	0.10999D-01	0.50531D 00
150.	0.55299D-02	0.73611D-02	0.75261D 00
165.	0.55313D-02	0.59264D-02	0.93370D 00
180.	0.55318D-02	0.55318D-02	0.10000D 01

ALPHA = 0.20

THETA	RHO S	RHO P	RHO U
0.	0.55044D-02	0.55044D-02	0.10000D 01
5.	0.55044D-02	0.55464D-02	0.99246D 00
10.	0.55045D-02	0.56752D-02	0.97009D 00
20.	0.55049D-02	0.62319D-02	0.88399D 00
30.	0.55058D-02	0.73332D-02	0.75216D 00
40.	0.55071D-02	0.93603D-02	0.59060D 00
45.	0.55080D-02	0.10973D-01	0.50471D 00
50.	0.55091D-02	0.13254D-01	0.41886D 00
60.	0.55119D-02	0.21739D-01	0.25764D 00
70.	0.55156D-02	0.45390D-01	0.12633D 00
80.	0.55202D-02	0.15592D 00	0.40698D-01
90.	0.55257D-02	0.10023D 01	0.10978D-01
100.	0.55320D-02	0.15618D 00	0.40727D-01
110.	0.55388D-02	0.45569D-01	0.12638D 00
120.	0.55458D-02	0.21870D-01	0.25770D 00
130.	0.55528D-02	0.13358D-01	0.41892D 00
135.	0.55560D-02	0.11068D-01	0.50476D 00
150.	0.55646D-02	0.74115D-02	0.75219D 00
165.	0.55704D-02	0.59691D-02	0.93356D 00
180.	0.55724D-02	0.55724D-02	0.10000D 01

ALPHA = 0.40

THETA	RHO S	RHO P	RHO U
0.	0.54621D-02	0.54621D-02	0.10000D 01
5.	0.54623D-02	0.55043D-02	0.99239D 00
10.	0.54627D-02	0.56336D-02	0.96982D 00
20.	0.54644D-02	0.61927D-02	0.88303D 00
30.	0.54676D-02	0.72994D-02	0.75041D 00
40.	0.54729D-02	0.93382D-02	0.58832D 00
45.	0.54765D-02	0.10961D-01	0.50235D 00
50.	0.54808D-02	0.13259D-01	0.41656D 00
60.	0.54919D-02	0.21815D-01	0.25583D 00
70.	0.55067D-02	0.45699D-01	0.12532D 00
80.	0.55254D-02	0.15747D 00	0.40391D-01
90.	0.55477D-02	0.10093D 01	0.10984D-01
100.	0.55733D-02	0.15853D 00	0.40505D-01
110.	0.56013D-02	0.46441D-01	0.12551D 00
120.	0.56303D-02	0.22355D-01	0.25605D 00
130.	0.56589D-02	0.13687D-01	0.41675D 00
135.	0.56725D-02	0.11352D-01	0.50251D 00
150.	0.57081D-02	0.76203D-02	0.75049D 00
165.	0.57319D-02	0.61461D-02	0.93299D 00
180.	0.57403D-02	0.57403D-02	0.10000D 01

ALPHA = 0.60

THETA	RHO S	RHO P	RHO U
0.	0.53917D-02	0.53917D-02	0.10000D 01
5.	0.53919D-02	0.54341D-02	0.99228D 00
10.	0.53928D-02	0.55642D-02	0.96936D 00
20.	0.53965D-02	0.61270D-02	0.88142D 00
30.	0.54036D-02	0.72424D-02	0.74748D 00
40.	0.54153D-02	0.93008D-02	0.58449D 00
45.	0.54233D-02	0.10942D-01	0.49837D 00
50.	0.54330D-02	0.13268D-01	0.41267D 00
60.	0.54581D-02	0.21947D-01	0.25277D 00
70.	0.54916D-02	0.46241D-01	0.12357D 00
80.	0.55343D-02	0.16019D 00	0.39863D-01
90.	0.55859D-02	0.10214D 01	0.10993D-01
100.	0.56454D-02	0.16270D 00	0.40118D-01
110.	0.57109D-02	0.48004D-01	0.12397D 00
120.	0.57793D-02	0.23227D-01	0.25314D 00
130.	0.58472D-02	0.14280D-01	0.41290D 00
135.	0.58796D-02	0.11865D-01	0.49851D 00
150.	0.59649D-02	0.79967D-02	0.74743D 00
165.	0.60223D-02	0.64648D-02	0.93196D 00
180.	0.60426D-02	0.60426D-02	0.10000D 01

ALPHA = 0.80

THETA	RHO S	RHO P	RHO U
0.	0.52929D-02	0.52929D-02	0.10000D 01
5.	0.52934D-02	0.53357D-02	0.99211D 00
10.	0.52949D-02	0.54667D-02	0.96873D 00
20.	0.53012D-02	0.60342D-02	0.87916D 00
30.	0.53134D-02	0.71612D-02	0.74333D 00
40.	0.53336D-02	0.92468D-02	0.57905D 00
45.	0.53476D-02	0.10914D-01	0.49271D 00
50.	0.53647D-02	0.13281D-01	0.40712D 00
60.	0.54092D-02	0.22143D-01	0.24835D 00
70.	0.54697D-02	0.47058D-01	0.12104D 00
80.	0.55474D-02	0.16434D 00	0.39087D-01
90.	0.56426D-02	0.10398D 01	0.11007D-01
100.	0.57536D-02	0.16916D 00	0.39539D-01
110.	0.58770D-02	0.50457D-01	0.12164D 00
120.	0.60074D-02	0.24603D-01	0.24869D 00
130.	0.61380D-02	0.15219D-01	0.40695D 00
135.	0.62009D-02	0.12677D-01	0.49230D 00
150.	0.63674D-02	0.85929D-02	0.74264D 00
165.	0.64804D-02	0.69690D-02	0.93034D 00
180.	0.65205D-02	0.65205D-02	0.10000D 01

ALPHA = 1.00

THETA	RHO S	RHO P	RHO U
0.	0.51659D-02	0.51659D-02	0.10000D 01
5.	0.51665D-02	0.52089D-02	0.99190D 00
10.	0.51687D-02	0.53409D-02	0.96791D 00
20.	0.51779D-02	0.59136D-02	0.87624D 00
30.	0.51960D-02	0.70542D-02	0.73795D 00
40.	0.52266D-02	0.91745D-02	0.57192D 00
45.	0.52480D-02	0.10876D-01	0.48525D 00
50.	0.52743D-02	0.13299D-01	0.39977D 00
60.	0.53440D-02	0.22416D-01	0.24245D 00
70.	0.54400D-02	0.48222D-01	0.11761D 00
80.	0.55655D-02	0.17036D 00	0.38022D-01
90.	0.57216D-02	0.10658D 01	0.11027D-01
100.	0.59066D-02	0.17877D 00	0.38719D-01
110.	0.61155D-02	0.54168D-01	0.11829D 00
120.	0.63398D-02	0.26705D-01	0.24221D 00
130.	0.65676D-02	0.16659D-01	0.39818D 00
135.	0.66784D-02	0.13924D-01	0.48309D 00
150.	0.69752D-02	0.95084D-02	0.73542D 00
165.	0.71792D-02	0.77416D-02	0.92787D 00
180.	0.72521D-02	0.72521D-02	0.10000D 01

P = 10.00   M = 1.40

ALPHA = 0.10

THETA	RHO S	RHO P	RHO U
0.	0.97669D-02	0.97669D-02	0.10000D 01
5.	0.97669D-02	0.98410D-02	0.99255D 00
10.	0.97669D-02	0.10068D-01	0.97041D 00
20.	0.97671D-02	0.11048D-01	0.88521D 00
30.	0.97673D-02	0.12983D-01	0.75470D 00
40.	0.97678D-02	0.16537D-01	0.59464D 00
45.	0.97682D-02	0.19355D-01	0.50949D 00
50.	0.97686D-02	0.23329D-01	0.42435D 00
60.	0.97696D-02	0.37987D-01	0.26437D 00
70.	0.97711D-02	0.77843D-01	0.13398D 00
80.	0.97729D-02	0.24677D 00	0.48899D-01
90.	0.97751D-02	0.10006D 01	0.19356D-01
100.	0.97777D-02	0.24684D 00	0.48911D-01
110.	0.97805D-02	0.77909D-01	0.13401D 00
120.	0.97834D-02	0.38038D-01	0.26440D 00
130.	0.97863D-02	0.23371D-01	0.42438D 00
135.	0.97877D-02	0.19393D-01	0.50951D 00
150.	0.97913D-02	0.13015D-01	0.75471D 00
165.	0.97937D-02	0.10490D-01	0.93427D 00
180.	0.97945D-02	0.97945D-02	0.10000D 01

ALPHA = 0.20

THETA	RHO S	RHO P	RHO U
0.	0.97516D-02	0.97516D-02	0.10000D 01
5.	0.97517D-02	0.98258D-02	0.99253D 00
10.	0.97518D-02	0.10053D-01	0.97034D 00
20.	0.97524D-02	0.11034D-01	0.88498D 00
30.	0.97535D-02	0.12972D-01	0.75428D 00
40.	0.97554D-02	0.16531D-01	0.59409D 00
45.	0.97567D-02	0.19354D-01	0.50892D 00
50.	0.97583D-02	0.23336D-01	0.42379D 00
60.	0.97626D-02	0.38025D-01	0.26393D 00
70.	0.97684D-02	0.77978D-01	0.13373D 00
80.	0.97758D-02	0.24736D 00	0.48819D-01
90.	0.97847D-02	0.10023D 01	0.19358D-01
100.	0.97951D-02	0.24764D 00	0.48870D-01
110.	0.98063D-02	0.78244D-01	0.13382D 00
120.	0.98181D-02	0.38230D-01	0.26404D 00
130.	0.98297D-02	0.23502D-01	0.42390D 00
135.	0.98352D-02	0.19507D-01	0.50902D 00
150.	0.98497D-02	0.13099D-01	0.75434D 00
165.	0.98594D-02	0.10562D-01	0.93414D 00
180.	0.98628D-02	0.98628D-02	0.10000D 01

ALPHA = 0.40

THETA	RHO S	RHO P	RHO U
0.	0.96906D-02	0.96906D-02	0.10000D 01
5.	0.96908D-02	0.97651D-02	0.99246D 00
10.	0.96913D-02	0.99932D-02	0.97008D 00
20.	0.96935D-02	0.10979D-01	0.88405D 00
30.	0.96979D-02	0.12927D-01	0.75259D 00
40.	0.97055D-02	0.16508D-01	0.59188D 00
45.	0.97107D-02	0.19350D-01	0.50663D 00
50.	0.97172D-02	0.23362D-01	0.42156D 00
60.	0.97343D-02	0.38177D-01	0.26216D 00
70.	0.97576D-02	0.78531D-01	0.13271D 00
80.	0.97875D-02	0.24977D 00	0.48499D-01
90.	0.98238D-02	0.10092D 01	0.19367D-01
100.	0.98658D-02	0.25092D 00	0.48704D-01
110.	0.99120D-02	0.79627D-01	0.13307D 00
120.	0.99602D-02	0.39023D-01	0.26259D 00
130.	0.10008D-01	0.24048D-01	0.42196D 00
135.	0.10031D-01	0.19980D-01	0.50700D 00
150.	0.10091D-01	0.13449D-01	0.75279D 00
165.	0.10131D-01	0.10859D-01	0.93362D 00
180.	0.10145D-01	0.10145D-01	0.10000D 01

ALPHA = 0.60

THETA	RHO S	RHO P	RHO U
0.	0.95889D-02	0.95889D-02	0.10000D 01
5.	0.95893D-02	0.96639D-02	0.99235D 00
10.	0.95904D-02	0.98936D-02	0.96965D 00
20.	0.95952D-02	0.10887D-01	0.88251D 00
30.	0.96049D-02	0.12852D-01	0.74976D 00
40.	0.96215D-02	0.16469D-01	0.58818D 00
45.	0.96333D-02	0.19345D-01	0.50278D 00
50.	0.96478D-02	0.23408D-01	0.41778D 00
60.	0.96863D-02	0.38441D-01	0.25916D 00
70.	0.97392D-02	0.79494D-01	0.13098D 00
80.	0.98075D-02	0.25401D 00	0.47948D-01
90.	0.98912D-02	0.10214D 01	0.19384D-01
100.	0.99887D-02	0.25672D 00	0.48415D-01
110.	0.10097D-01	0.82091D-01	0.13176D 00
120.	0.10210D-01	0.40441D-01	0.26002D 00
130.	0.10323D-01	0.25025D-01	0.41852D 00
135.	0.10377D-01	0.20827D-01	0.50341D 00
150.	0.10520D-01	0.14076D-01	0.75003D 00
165.	0.10616D-01	0.11391D-01	0.93269D 00
180.	0.10650D-01	0.10650D-01	0.10000D 01

ALPHA = 0.80

THETA	RHO S	RHO P	RHO U
0.	0.94465D-02	0.94465D-02	0.10000D 01
5.	0.94471D-02	0.95221D-02	0.99219D 00
10.	0.94489D-02	0.97538D-02	0.96904D 00
20.	0.94570D-02	0.10756D-01	0.88033D 00
30.	0.94736D-02	0.12744D-01	0.74577D 00
40.	0.95025D-02	0.16413D-01	0.58292D 00
45.	0.95232D-02	0.19336D-01	0.49729D 00
50.	0.95488D-02	0.23474D-01	0.41239D 00
60.	0.96173D-02	0.38831D-01	0.25484D 00
70.	0.97124D-02	0.80941D-01	0.12846D 00
80.	0.98368D-02	0.26045D 00	0.47142D-01
90.	0.99909D-02	0.10396D 01	0.19408D-01
100.	0.10172D-01	0.26560D 00	0.47983D-01
110.	0.10375D-01	0.85920D-01	0.12978D 00
120.	0.10590D-01	0.42661D-01	0.25613D 00
130.	0.10807D-01	0.26558D-01	0.41325D 00
135.	0.10911D-01	0.22159D-01	0.49789D 00
150.	0.11188D-01	0.15060D-01	0.74574D 00
165.	0.11376D-01	0.12227D-01	0.93123D 00
180.	0.11443D-01	0.11443D-01	0.10000D 01

ALPHA = 1.00

THETA	RHO S	RHO P	RHO U
0.	0.92632D-02	0.92632D-02	0.10000D 01
5.	0.92640D-02	0.93395D-02	0.99199D 00
10.	0.92667D-02	0.95734D-02	0.96825D 00
20.	0.92785D-02	0.10587D-01	0.87753D 00
30.	0.93031D-02	0.12603D-01	0.74058D 00
40.	0.93469D-02	0.16338D-01	0.57604D 00
45.	0.93785D-02	0.19325D-01	0.49009D 00
50.	0.94181D-02	0.23565D-01	0.40527D 00
60.	0.95253D-02	0.39374D-01	0.24907D 00
70.	0.96763D-02	0.82994D-01	0.12506D 00
80.	0.98768D-02	0.26973D 00	0.46040D-01
90.	0.10129D-01	0.10652D 01	0.19441D-01
100.	0.10430D-01	0.27858D 00	0.47376D-01
110.	0.10772D-01	0.91630D-01	0.12696D 00
120.	0.11140D-01	0.46004D-01	0.25050D 00
130.	0.11514D-01	0.28880D-01	0.40555D 00
135.	0.11697D-01	0.24177D-01	0.48977D 00
150.	0.12185D-01	0.16552D-01	0.73935D 00
165.	0.12522D-01	0.13491D-01	0.92904D 00
180.	0.12642D-01	0.12642D-01	0.10000D 01

ALPHA = 0.10

THETA	RHO S	RHO P	RHO U
0.	0.11653D-03	0.11653D-03	0.10000D 01
5.	0.11653D-03	0.11742D-03	0.99240D 00
10.	0.11653D-03	0.12016D-03	0.96983D 00
20.	0.11654D-03	0.13199D-03	0.88296D 00
30.	0.11655D-03	0.15542D-03	0.74990D 00
40.	0.11656D-03	0.19869D-03	0.58672D 00
45.	0.11657D-03	0.23321D-03	0.49991D 00
50.	0.11658D-03	0.28225D-03	0.41311D 00
60.	0.11661D-03	0.46656D-03	0.25002D 00
70.	0.11664D-03	0.99694D-03	0.11710D 00
80.	0.11668D-03	0.38578D-02	0.30357D-01
90.	0.11672D-03	0.10005D 01	0.23334D-03
100.	0.11676D-03	0.38611D-02	0.30354D-01
110.	0.11681D-03	0.99845D-03	0.11709D 00
120.	0.11686D-03	0.46757D-03	0.25001D 00
130.	0.11690D-03	0.28304D-03	0.41311D 00
135.	0.11693D-03	0.23393D-03	0.49990D 00
150.	0.11698D-03	0.15600D-03	0.74990D 00
165.	0.11702D-03	0.12543D-03	0.93298D 00
180.	0.11703D-03	0.11703D-03	0.10000D 01

ALPHA = 0.20

THETA	RHO S	RHO P	RHO U
0.	0.11618D-03	0.11618D-03	0.10000D 01
5.	0.11618D-03	0.11707D-03	0.99238D 00
10.	0.11618D-03	0.11981D-03	0.96976D 00
20.	0.11621D-03	0.13165D-03	0.88271D 00
30.	0.11625D-03	0.15512D-03	0.74944D 00
40.	0.11630D-03	0.19845D-03	0.58612D 00
45.	0.11634D-03	0.23304D-03	0.49929D 00
50.	0.11638D-03	0.28218D-03	0.41251D 00
60.	0.11648D-03	0.46692D-03	0.24956D 00
70.	0.11661D-03	0.99883D-03	0.11685D 00
80.	0.11675D-03	0.38693D-02	0.30287D-01
90.	0.11692D-03	0.10022D 01	0.23356D-03
100.	0.11710D-03	0.38824D-02	0.30276D-01
110.	0.11730D-03	0.10049D-02	0.11683D 00
120.	0.11749D-03	0.47101D-03	0.24953D 00
130.	0.11768D-03	0.28534D-03	0.41248D 00
135.	0.11777D-03	0.23591D-03	0.49926D 00
150.	0.11799D-03	0.15745D-03	0.74942D 00
165.	0.11814D-03	0.12665D-03	0.93281D 00
180.	0.11819D-03	0.11819D-03	0.10000D 01

ALPHA = 0.40

THETA	RHO S	RHO P	RHO U
0.	0.11475D-03	0.11475D-03	0.10000D 01
5.	0.11476D-03	0.11565D-03	0.99231D 00
10.	0.11478D-03	0.11840D-03	0.96947D 00
20.	0.11487D-03	0.13029D-03	0.88168D 00
30.	0.11503D-03	0.15388D-03	0.74757D 00
40.	0.11526D-03	0.19748D-03	0.58370D 00
45.	0.11541D-03	0.23233D-03	0.49680D 00
50.	0.11557D-03	0.28187D-03	0.41009D 00
60.	0.11598D-03	0.46841D-03	0.24769D 00
70.	0.11648D-03	0.10066D-02	0.11582D 00
80.	0.11707D-03	0.39167D-02	0.30005D-01
90.	0.11775D-03	0.10090D 01	0.23443D-03
100.	0.11850D-03	0.39710D-02	0.29957D-01
110.	0.11930D-03	0.10318D-02	0.11572D 00
120.	0.12010D-03	0.48537D-03	0.24753D 00
130.	0.12088D-03	0.29495D-03	0.40989D 00
135.	0.12124D-03	0.24418D-03	0.49659D 00
150.	0.12219D-03	0.16349D-03	0.74741D 00
165.	0.12282D-03	0.13176D-03	0.93214D 00
180.	0.12304D-03	0.12304D-03	0.10000D 01

ALPHA = 0.60

THETA	RHO S	RHO P	RHO U
0.	0.11239D-03	0.11239D-03	0.10000D 01
5.	0.11240D-03	0.11329D-03	0.99218D 00
10.	0.11245D-03	0.11605D-03	0.96898D 00
20.	0.11265D-03	0.12802D-03	0.87997D 00
30.	0.11299D-03	0.15179D-03	0.74444D 00
40.	0.11350D-03	0.19584D-03	0.57964D 00
45.	0.11383D-03	0.23111D-03	0.49259D 00
50.	0.11420D-03	0.28134D-03	0.40600D 00
60.	0.11512D-03	0.47099D-03	0.24450D 00
70.	0.11626D-03	0.10202D-02	0.11406D 00
80.	0.11763D-03	0.40006D-02	0.29516D-01
90.	0.11921D-03	0.10208D 01	0.23596D-03
100.	0.12096D-03	0.41316D-02	0.29395D-01
110.	0.12284D-03	0.10809D-02	0.11376D 00
120.	0.12476D-03	0.51161D-03	0.24396D 00
130.	0.12664D-03	0.31255D-03	0.40524D 00
135.	0.12752D-03	0.25934D-03	0.49177D 00
150.	0.12983D-03	0.17457D-03	0.74376D 00
165.	0.13137D-03	0.14112D-03	0.93092D 00
180.	0.13191D-03	0.13191D-03	0.10000D 01

ALPHA = 0.80

THETA	RHO S	RHO P	RHO U
0.	0.10908D-03	0.10908D-03	0.10000D 01
5.	0.10911D-03	0.10999D-03	0.99201D 00
10.	0.10919D-03	0.11277D-03	0.96830D 00
20.	0.10953D-03	0.12481D-03	0.87756D 00
30.	0.11012D-03	0.14881D-03	0.74003D 00
40.	0.11101D-03	0.19346D-03	0.57386D 00
45.	0.11158D-03	0.22934D-03	0.48659D 00
50.	0.11224D-03	0.28056D-03	0.40013D 00
60.	0.11387D-03	0.47486D-03	0.23989D 00
70.	0.11593D-03	0.10408D-02	0.11149D 00
80.	0.11844D-03	0.41304D-02	0.28791D-01
90.	0.12139D-03	0.10384D 01	0.23825D-03
100.	0.12471D-03	0.43880D-02	0.28541D-01
110.	0.12830D-03	0.11599D-02	0.11073D 00
120.	0.13204D-03	0.55415D-03	0.23838D 00
130.	0.13573D-03	0.34119D-03	0.39789D 00
135.	0.13749D-03	0.28404D-03	0.48413D 00
150.	0.14213D-03	0.19262D-03	0.73789D 00
165.	0.14525D-03	0.15637D-03	0.92893D 00
180.	0.14636D-03	0.14636D-03	0.10000D 01

ALPHA = 1.00

THETA	RHO S	RHO P	RHO U
0.	0.10484D-03	0.10484D-03	0.10000D 01
5.	0.10488D-03	0.10575D-03	0.99178D 00
10.	0.10500D-03	0.10853D-03	0.96743D 00
20.	0.10550D-03	0.12065D-03	0.87445D 00
30.	0.10639D-03	0.14489D-03	0.73431D 00
40.	0.10774D-03	0.19027D-03	0.56629D 00
45.	0.10862D-03	0.22693D-03	0.47868D 00
50.	0.10964D-03	0.27950D-03	0.39236D 00
60.	0.11220D-03	0.48029D-03	0.23369D 00
70.	0.11549D-03	0.10707D-02	0.10797D 00
80.	0.11958D-03	0.43221D-02	0.27784D-01
90.	0.12447D-03	0.10634D 01	0.24150D-03
100.	0.13011D-03	0.47872D-02	0.27306D-01
110.	0.13636D-03	0.12845D-02	0.10628D 00
120.	0.14297D-03	0.62192D-03	0.23000D 00
130.	0.14963D-03	0.38709D-03	0.38665D 00
135.	0.15286D-03	0.32369D-03	0.47233D 00
150.	0.16148D-03	0.22163D-03	0.72864D 00
165.	0.16740D-03	0.18083D-03	0.92576D 00
180.	0.16951D-03	0.16951D-03	0.10000D 01

ALPHA = 0.10

THETA	RHO S	RHO P	RHO U
0.	0.46952D-03	0.46952D-03	0.10000D 01
5.	0.46952D-03	0.47311D-03	0.99240D 00
10.	0.46952D-03	0.48413D-03	0.96985D 00
20.	0.46954D-03	0.53176D-03	0.88305D 00
30.	0.46958D-03	0.62614D-03	0.75008D 00
40.	0.46964D-03	0.80030D-03	0.58702D 00
45.	0.46967D-03	0.93928D-03	0.50027D 00
50.	0.46971D-03	0.11366D-02	0.41353D 00
60.	0.46981D-03	0.18777D-02	0.25056D 00
70.	0.46992D-03	0.40058D-02	0.11773D 00
80.	0.47007D-03	0.15368D-01	0.31043D-01
90.	0.47023D-03	0.10005D 01	0.93975D-03
100.	0.47040D-03	0.15381D-01	0.31039D-01
110.	0.47059D-03	0.40117D-02	0.11772D 00
120.	0.47077D-03	0.18816D-02	0.25055D 00
130.	0.47095D-03	0.11396D-02	0.41352D 00
135.	0.47104D-03	0.94203D-03	0.50026D 00
150.	0.47126D-03	0.62838D-03	0.75008D 00
165.	0.47140D-03	0.50526D-03	0.93302D 00
180.	0.47145D-03	0.47145D-03	0.10000D 01

ALPHA = 0.20

THETA	RHO S	RHO P	RHO U
0.	0.46827D-03	0.46827D-03	0.10000D 01
5.	0.46827D-03	0.47187D-03	0.99239D 00
10.	0.46829D-03	0.48289D-03	0.96978D 00
20.	0.46838D-03	0.53059D-03	0.88280D 00
30.	0.46853D-03	0.62510D-03	0.74963D 00
40.	0.46874D-03	0.79957D-03	0.58644D 00
45.	0.46888D-03	0.93882D-03	0.49967D 00
50.	0.46904D-03	0.11366D-02	0.41295D 00
60.	0.46942D-03	0.18795D-02	0.25011D 00
70.	0.46990D-03	0.40138D-02	0.11749D 00
80.	0.47046D-03	0.15413D-01	0.30979D-01
90.	0.47110D-03	0.10022D 01	0.94073D-03
100.	0.47181D-03	0.15468D-01	0.30960D-01
110.	0.47256D-03	0.40378D-02	0.11745D 00
120.	0.47331D-03	0.18955D-02	0.25007D 00
130.	0.47404D-03	0.11488D-02	0.41291D 00
135.	0.47439D-03	0.94994D-03	0.49962D 00
150.	0.47527D-03	0.63413D-03	0.74960D 00
165.	0.47585D-03	0.51012D-03	0.93287D 00
180.	0.47606D-03	0.47606D-03	0.10000D 01

ALPHA = 0.40

THETA	RHO S	RHO P	RHO U
0.	0.46327D-03	0.46327D-03	0.10000D 01
5.	0.46330D-03	0.46689D-03	0.99231D 00
10.	0.46338D-03	0.47796D-03	0.96950D 00
20.	0.46371D-03	0.52590D-03	0.88180D 00
30.	0.46429D-03	0.62096D-03	0.74782D 00
40.	0.46516D-03	0.79663D-03	0.58409D 00
45.	0.46570D-03	0.93699D-03	0.49726D 00
50.	0.46634D-03	0.11365D-02	0.41062D 00
60.	0.46788D-03	0.18869D-02	0.24831D 00
70.	0.46979D-03	0.40466D-02	0.11651D 00
80.	0.47207D-03	0.15600D-01	0.30718D-01
90.	0.47469D-03	0.10090D 01	0.94472D-03
100.	0.47759D-03	0.15827D-01	0.30639D-01
110.	0.48066D-03	0.41463D-02	0.11635D 00
120.	0.48379D-03	0.19530D-02	0.24808D 00
130.	0.48682D-03	0.11872D-02	0.41035D 00
135.	0.48825D-03	0.98289D-03	0.49699D 00
150.	0.49195D-03	0.65812D-03	0.74763D 00
165.	0.49440D-03	0.53037D-03	0.93221D 00
180.	0.49525D-03	0.49525D-03	0.10000D 01

ALPHA = 0.60

THETA	RHO S	RHO P	RHO U
0.	0.45496D-03	0.45496D-03	0.10000D 01
5.	0.45502D-03	0.45860D-03	0.99219D 00
10.	0.45520D-03	0.46976D-03	0.96903D 00
20.	0.45593D-03	0.51805D-03	0.88014D 00
30.	0.45721D-03	0.61397D-03	0.74479D 00
40.	0.45913D-03	0.79164D-03	0.58016D 00
45.	0.46035D-03	0.93386D-03	0.49319D 00
50.	0.46177D-03	0.11363D-02	0.40666D 00
60.	0.46524D-03	0.18997D-02	0.24525D 00
70.	0.46960D-03	0.41043D-02	0.11483D 00
80.	0.47484D-03	0.15931D-01	0.30267D-01
90.	0.48093D-03	0.10207D 01	0.95166D-03
100.	0.48772D-03	0.16476D-01	0.30075D-01
110.	0.49499D-03	0.43434D-02	0.11440D 00
120.	0.50245D-03	0.20578D-02	0.24456D 00
130.	0.50974D-03	0.12572D-02	0.40577D 00
135.	0.51319D-03	0.10431D-02	0.49225D 00
150.	0.52219D-03	0.70195D-03	0.74404D 00
165.	0.52819D-03	0.56736D-03	0.93100D 00
180.	0.53030D-03	0.53030D-03	0.10000D 01

ALPHA = 0.80

THETA	RHO S	RHO P	RHO U
0.	0.44337D-03	0.44337D-03	0.10000D 01
5.	0.44347D-03	0.44704D-03	0.99202D 00
10.	0.44377D-03	0.45827D-03	0.96837D 00
20.	0.44503D-03	0.50700D-03	0.87781D 00
30.	0.44724D-03	0.60404D-03	0.74053D 00
40.	0.45058D-03	0.78445D-03	0.57459D 00
45.	0.45274D-03	0.92931D-03	0.48741D 00
50.	0.45525D-03	0.11360D-02	0.40102D 00
60.	0.46144D-03	0.19188D-02	0.24083D 00
70.	0.46932D-03	0.41914D-02	0.11239D 00
80.	0.47894D-03	0.16438D-01	0.29601D-01
90.	0.49027D-03	0.10381D 01	0.96205D-03
100.	0.50308D-03	0.17510D-01	0.29220D-01
110.	0.51700D-03	0.46595D-02	0.11141D 00
120.	0.53146D-03	0.22268D-02	0.23907D 00
130.	0.54577D-03	0.13705D-02	0.39857D 00
135.	0.55260D-03	0.11406D-02	0.48477D 00
150.	0.57059D-03	0.77299D-03	0.73831D 00
165.	0.58273D-03	0.62724D-03	0.92907D 00
180.	0.58701D-03	0.58701D-03	0.10000D 01

ALPHA = 1.00

THETA	RHO S	RHO P	RHO U
0.	0.42853D-03	0.42853D-03	0.10000D 01
5.	0.42867D-03	0.43222D-03	0.99181D 00
10.	0.42913D-03	0.44353D-03	0.96753D 00
20.	0.43099D-03	0.49270D-03	0.87481D 00
30.	0.43432D-03	0.59100D-03	0.73501D 00
40.	0.43941D-03	0.77483D-03	0.56730D 00
45.	0.44273D-03	0.92317D-03	0.47980D 00
50.	0.44662D-03	0.11356D-02	0.39355D 00
60.	0.45635D-03	0.19456D-02	0.23491D 00
70.	0.46894D-03	0.43164D-02	0.10906D 00
80.	0.48462D-03	0.17181D-01	0.28678D-01
90.	0.50343D-03	0.10626D 01	0.97669D-03
100.	0.52514D-03	0.19109D-01	0.27992D-01
110.	0.54921D-03	0.51545D-02	0.10704D 00
120.	0.57473D-03	0.24939D-02	0.23089D 00
130.	0.60043D-03	0.15504D-02	0.38764D 00
135.	0.61289D-03	0.12958D-02	0.47332D 00
150.	0.64614D-03	0.88611D-03	0.72936D 00
165.	0.66896D-03	0.72246D-03	0.92600D 00
180.	0.67710D-03	0.67710D-03	0.10000D 01

ALPHA = 0.10

THETA	RHO S	RHO P	RHO U
0.	0.18916D-02	0.18916D-02	0.10000D 01
5.	0.18916D-02	0.19061D-02	0.99243D 00
10.	0.18916D-02	0.19504D-02	0.96994D 00
20.	0.18917D-02	0.21419D-02	0.88339D 00
30.	0.18918D-02	0.25213D-02	0.75080D 00
40.	0.18920D-02	0.32209D-02	0.58820D 00
45.	0.18921D-02	0.37785D-02	0.50170D 00
50.	0.18923D-02	0.45696D-02	0.41521D 00
60.	0.18926D-02	0.75319D-02	0.25270D 00
70.	0.18931D-02	0.15965D-01	0.12024D 00
80.	0.18936D-02	0.59228D-01	0.33800D-01
90.	0.18942D-02	0.10005D 01	0.37801D-02
100.	0.18948D-02	0.59282D-01	0.33794D-01
110.	0.18955D-02	0.15988D-01	0.12023D 00
120.	0.18962D-02	0.75468D-02	0.25268D 00
130.	0.18969D-02	0.45809D-02	0.41520D 00
135.	0.18972D-02	0.37889D-02	0.50169D 00
150.	0.18981D-02	0.25297D-02	0.75079D 00
165.	0.18986D-02	0.20348D-02	0.93322D 00
180.	0.18988D-02	0.18988D-02	0.10000D 01

ALPHA = 0.20

THETA	RHO S	RHO P	RHO U
0.	0.18879D-02	0.18879D-02	0.10000D 01
5.	0.18879D-02	0.19024D-02	0.99241D 00
10.	0.18880D-02	0.19467D-02	0.96987D 00
20.	0.18883D-02	0.21386D-02	0.88315D 00
30.	0.18888D-02	0.25187D-02	0.75038D 00
40.	0.18895D-02	0.32197D-02	0.58765D 00
45.	0.18900D-02	0.37786D-02	0.50114D 00
50.	0.18906D-02	0.45715D-02	0.41467D 00
60.	0.18920D-02	0.75418D-02	0.25228D 00
70.	0.18937D-02	0.16000D-01	0.12002D 00
80.	0.18958D-02	0.59403D-01	0.33746D-01
90.	0.18982D-02	0.10022D 01	0.37850D-02
100.	0.19008D-02	0.59619D-01	0.33719D-01
110.	0.19036D-02	0.16092D-01	0.11997D 00
120.	0.19065D-02	0.76018D-02	0.25222D 00
130.	0.19092D-02	0.46175D-02	0.41460D 00
135.	0.19105D-02	0.38202D-02	0.50108D 00
150.	0.19139D-02	0.25524D-02	0.75034D 00
165.	0.19162D-02	0.20539D-02	0.93306D 00
180.	0.19169D-02	0.19169D-02	0.10000D 01

ALPHA = 0.40

THETA	RHO S	RHO P	RHO U
0.	0.18731D-02	0.18731D-02	0.10000D 01
5.	0.18732D-02	0.18877D-02	0.99234D 00
10.	0.18735D-02	0.19323D-02	0.96960D 00
20.	0.18746D-02	0.21254D-02	0.88221D 00
30.	0.18766D-02	0.25082D-02	0.74867D 00
40.	0.18796D-02	0.32148D-02	0.58545D 00
45.	0.18816D-02	0.37788D-02	0.49887D 00
50.	0.18838D-02	0.45794D-02	0.41247D 00
60.	0.18894D-02	0.75821D-02	0.25060D 00
70.	0.18963D-02	0.16145D-01	0.11912D 00
80.	0.19047D-02	0.60116D-01	0.33525D-01
90.	0.19145D-02	0.10089D 01	0.38047D-02
100.	0.19253D-02	0.61015D-01	0.33415D-01
110.	0.19368D-02	0.16525D-01	0.11891D 00
120.	0.19486D-02	0.78305D-02	0.25031D 00
130.	0.19601D-02	0.47692D-02	0.41215D 00
135.	0.19655D-02	0.39504D-02	0.49855D 00
150.	0.19796D-02	0.26468D-02	0.74844D 00
165.	0.19890D-02	0.21334D-02	0.93243D 00
180.	0.19923D-02	0.19923D-02	0.10000D 01

ALPHA = 0.60

THETA	RHO S	RHO P	RHO U
0.	0.18485D-02	0.18485D-02	0.10000D 01
5.	0.18487D-02	0.18632D-02	0.99223D 00
10.	0.18493D-02	0.19083D-02	0.96916D 00
20.	0.18518D-02	0.21034D-02	0.88064D 00
30.	0.18563D-02	0.24905D-02	0.74582D 00
40.	0.18630D-02	0.32067D-02	0.58176D 00
45.	0.18674D-02	0.37792D-02	0.49506D 00
50.	0.18724D-02	0.45930D-02	0.40878D 00
60.	0.18849D-02	0.76516D-02	0.24776D 00
70.	0.19008D-02	0.16398D-01	0.11760D 00
80.	0.19201D-02	0.61366D-01	0.33146D-01
90.	0.19427D-02	0.10205D 01	0.38388D-02
100.	0.19680D-02	0.63523D-01	0.32884D-01
110.	0.19953D-02	0.17307D-01	0.11705D 00
120.	0.20234D-02	0.82446D-02	0.24694D 00
130.	0.20509D-02	0.50445D-02	0.40777D 00
135.	0.20639D-02	0.41866D-02	0.49403D 00
150.	0.20980D-02	0.28181D-02	0.74503D 00
165.	0.21208D-02	0.22777D-02	0.93129D 00
180.	0.21288D-02	0.21288D-02	0.10000D 01

ALPHA = 0.80

THETA	RHO S	RHO P	RHO U
0.	0.18144D-02	0.18144D-02	0.10000D 01
5.	0.18147D-02	0.18293D-02	0.99207D 00
10.	0.18158D-02	0.18748D-02	0.96855D 00
20.	0.18201D-02	0.20724D-02	0.87846D 00
30.	0.18278D-02	0.24655D-02	0.74183D 00
40.	0.18396D-02	0.31949D-02	0.57655D 00
45.	0.18472D-02	0.37797D-02	0.48967D 00
50.	0.18562D-02	0.46126D-02	0.40353D 00
60.	0.18786D-02	0.77544D-02	0.24368D 00
70.	0.19073D-02	0.16776D-01	0.11538D 00
80.	0.19427D-02	0.63261D-01	0.32588D-01
90.	0.19846D-02	0.10376D 01	0.38896D-02
100.	0.20323D-02	0.67475D-01	0.32086D-01
110.	0.20843D-02	0.18548D-01	0.11422D 00
120.	0.21385D-02	0.89058D-02	0.24174D 00
130.	0.21922D-02	0.54852D-02	0.40096D 00
135.	0.22179D-02	0.45652D-02	0.48696D 00
150.	0.22856D-02	0.30926D-02	0.73963D 00
165.	0.23313D-02	0.25086D-02	0.92946D 00
180.	0.23474D-02	0.23474D-02	0.10000D 01

ALPHA = 1.00

THETA	RHO S	RHO P	RHO U
0.	0.17709D-02	0.17709D-02	0.10000D 01
5.	0.17714D-02	0.17859D-02	0.99187D 00
10.	0.17729D-02	0.18321D-02	0.96776D 00
20.	0.17794D-02	0.20326D-02	0.87566D 00
30.	0.17910D-02	0.24328D-02	0.73668D 00
40.	0.18091D-02	0.31794D-02	0.56978D 00
45.	0.18209D-02	0.37804D-02	0.48261D 00
50.	0.18349D-02	0.46393D-02	0.39662D 00
60.	0.18701D-02	0.78966D-02	0.23825D 00
70.	0.19161D-02	0.17312D-01	0.11238D 00
80.	0.19737D-02	0.65981D-01	0.31824D-01
90.	0.20431D-02	0.10612D 01	0.39604D-02
100.	0.21236D-02	0.73496D-01	0.30952D-01
110.	0.22130D-02	0.20463D-01	0.11012D 00
120.	0.23079D-02	0.99334D-02	0.23410D 00
130.	0.24035D-02	0.61735D-02	0.39078D 00
135.	0.24498D-02	0.51572D-02	0.47631D 00
150.	0.25735D-02	0.35222D-02	0.73133D 00
165.	0.26583D-02	0.28694D-02	0.92663D 00
180.	0.26885D-02	0.26885D-02	0.10000D 01

ALPHA = 0.10

THETA	RHO S	RHO P	RHO U
0.	0.42488D-02	0.42488D-02	0.10000D 01
5.	0.42488D-02	0.42812D-02	0.99246D 00
10.	0.42489D-02	0.43805D-02	0.97008D 00
20.	0.42490D-02	0.48096D-02	0.88394D 00
30.	0.42492D-02	0.56587D-02	0.75198D 00
40.	0.42496D-02	0.72222D-02	0.59015D 00
45.	0.42499D-02	0.84667D-02	0.50406D 00
50.	0.42501D-02	0.10229D-01	0.41797D 00
60.	0.42508D-02	0.16798D-01	0.25622D 00
70.	0.42517D-02	0.35236D-01	0.12439D 00
80.	0.42528D-02	0.12413D 00	0.38349D-01
90.	0.42540D-02	0.10005D 01	0.84697D-02
100.	0.42554D-02	0.12424D 00	0.38343D-01
110.	0.42569D-02	0.35284D-01	0.12437D 00
120.	0.42584D-02	0.16829D-01	0.25620D 00
130.	0.42599D-02	0.10253D-01	0.41796D 00
135.	0.42605D-02	0.84884D-02	0.50404D 00
150.	0.42623D-02	0.56762D-02	0.75197D 00
165.	0.42635D-02	0.45685D-02	0.93353D 00
180.	0.42639D-02	0.42639D-02	0.10000D 01

ALPHA = 0.20

THETA	RHO S	RHO P	RHO U
0.	0.42431D-02	0.42431D-02	0.10000D 01
5.	0.42432D-02	0.42756D-02	0.99245D 00
10.	0.42433D-02	0.43750D-02	0.97002D 00
20.	0.42439D-02	0.48050D-02	0.88372D 00
30.	0.42448D-02	0.56558D-02	0.75158D 00
40.	0.42463D-02	0.72230D-02	0.58963D 00
45.	0.42473D-02	0.84706D-02	0.50352D 00
50.	0.42484D-02	0.10238D-01	0.41746D 00
60.	0.42512D-02	0.16825D-01	0.25583D 00
70.	0.42547D-02	0.35322D-01	0.12418D 00
80.	0.42590D-02	0.12450D 00	0.38306D-01
90.	0.42640D-02	0.10022D 01	0.84825D-02
100.	0.42696D-02	0.12494D 00	0.38280D-01
110.	0.42755D-02	0.35514D-01	0.12413D 00
120.	0.42816D-02	0.16951D-01	0.25577D 00
130.	0.42875D-02	0.10334D-01	0.41740D 00
135.	0.42903D-02	0.85577D-02	0.50346D 00
150.	0.42975D-02	0.57264D-02	0.75154D 00
165.	0.43023D-02	0.46107D-02	0.93339D 00
180.	0.43040D-02	0.43040D-02	0.10000D 01

ALPHA = 0.40

THETA	RHO S	RHO P	RHO U
0.	0.42205D-02	0.42205D-02	0.10000D 01
5.	0.42206D-02	0.42532D-02	0.99238D 00
10.	0.42212D-02	0.43533D-02	0.96977D 00
20.	0.42233D-02	0.47866D-02	0.88283D 00
30.	0.42272D-02	0.56445D-02	0.74997D 00
40.	0.42332D-02	0.72262D-02	0.58756D 00
45.	0.42370D-02	0.84863D-02	0.50139D 00
50.	0.42415D-02	0.10272D-01	0.41540D 00
60.	0.42527D-02	0.16937D-01	0.25427D 00
70.	0.42670D-02	0.35670D-01	0.12337D 00
80.	0.42843D-02	0.12597D 00	0.38131D-01
90.	0.43046D-02	0.10089D 01	0.85346D-02
100.	0.43274D-02	0.12781D 00	0.38021D-01
110.	0.43518D-02	0.36465D-01	0.12316D 00
120.	0.43768D-02	0.17457D-01	0.25398D 00
130.	0.44012D-02	0.10669D-01	0.41508D 00
135.	0.44128D-02	0.88454D-02	0.50108D 00
150.	0.44428D-02	0.59345D-02	0.74975D 00
165.	0.44628D-02	0.47859D-02	0.93279D 00
180.	0.44698D-02	0.44698D-02	0.10000D 01

ALPHA = 0.60

THETA	RHO S	RHO P	RHO U
0.	0.41830D-02	0.41830D-02	0.10000D 01
5.	0.41834D-02	0.42161D-02	0.99227D 00
10.	0.41845D-02	0.43174D-02	0.96935D 00
20.	0.41893D-02	0.47560D-02	0.88135D 00
30.	0.41979D-02	0.56255D-02	0.74729D 00
40.	0.42112D-02	0.72315D-02	0.58409D 00
45.	0.42198D-02	0.85129D-02	0.49782D 00
50.	0.42300D-02	0.10331D-01	0.41194D 00
60.	0.42553D-02	0.17128D-01	0.25163D 00
70.	0.42878D-02	0.36272D-01	0.12198D 00
80.	0.43277D-02	0.12854D 00	0.37831D-01
90.	0.43747D-02	0.10203D 01	0.86244D-02
100.	0.44278D-02	0.13292D 00	0.37572D-01
110.	0.44852D-02	0.38171D-01	0.12144D 00
120.	0.45446D-02	0.18367D-01	0.25084D 00
130.	0.46029D-02	0.11274D-01	0.41098D 00
135.	0.46306D-02	0.93641D-02	0.49684D 00
150.	0.47031D-02	0.63099D-02	0.74654D 00
165.	0.47516D-02	0.51016D-02	0.93172D 00
180.	0.47687D-02	0.47687D-02	0.10000D 01

ALPHA = 0.80

THETA	RHO S	RHO P	RHO U
0.	0.41310D-02	0.41310D-02	0.10000D 01
5.	0.41317D-02	0.41646D-02	0.99212D 00
10.	0.41337D-02	0.42675D-02	0.96877D 00
20.	0.41420D-02	0.47132D-02	0.87930D 00
30.	0.41570D-02	0.55988D-02	0.74354D 00
40.	0.41802D-02	0.72391D-02	0.57921D 00
45.	0.41955D-02	0.85510D-02	0.49278D 00
50.	0.42136D-02	0.10416D-01	0.40705D 00
60.	0.42590D-02	0.17408D-01	0.24786D 00
70.	0.43178D-02	0.37168D-01	0.11997D 00
80.	0.43909D-02	0.13239D 00	0.37393D-01
90.	0.44780D-02	0.10371D 01	0.87569D-02
100.	0.45777D-02	0.14088D 00	0.36903D-01
110.	0.46867D-02	0.40851D-01	0.11886D 00
120.	0.48007D-02	0.19804D-01	0.24603D 00
130.	0.49139D-02	0.12232D-01	0.40466D 00
135.	0.49681D-02	0.10186D-01	0.49027D 00
150.	0.51111D-02	0.69050D-02	0.74152D 00
165.	0.52077D-02	0.56017D-02	0.93002D 00
180.	0.52419D-02	0.52419D-02	0.10000D 01

ALPHA = 1.00

THETA	RHO S	RHO P	RHO U
0.	0.40651D-02	0.40651D-02	0.10000D 01
5.	0.40661D-02	0.40993D-02	0.99194D 00
10.	0.40691D-02	0.42041D-02	0.96803D 00
20.	0.40817D-02	0.46585D-02	0.87668D 00
30.	0.41045D-02	0.55642D-02	0.73874D 00
40.	0.41403D-02	0.72491D-02	0.57292D 00
45.	0.41640D-02	0.86017D-02	0.48624D 00
50.	0.41923D-02	0.10529D-01	0.40066D 00
60.	0.42638D-02	0.17792D-01	0.24287D 00
70.	0.43580D-02	0.38418D-01	0.11728D 00
80.	0.44769D-02	0.13781D 00	0.36798D-01
90.	0.46209D-02	0.10598D 01	0.89400D-02
100.	0.47884D-02	0.15276D 00	0.35961D-01
110.	0.49748D-02	0.44914D-01	0.11516D 00
120.	0.51728D-02	0.22001D-01	0.23906D 00
130.	0.53726D-02	0.13702D-01	0.39535D 00
135.	0.54694D-02	0.11450D-01	0.48052D 00
150.	0.57276D-02	0.78202D-02	0.73394D 00
165.	0.59047D-02	0.63697D-02	0.92743D 00
180.	0.59678D-02	0.59678D-02	0.10000D 01

ALPHA = 0.10

THETA	RHO S	RHO P	RHO U
0.	0.74813D-02	0.74813D-02	0.10000D 01
5.	0.74814D-02	0.75382D-02	0.99251D 00
10.	0.74814D-02	0.77124D-02	0.97027D 00
20.	0.74816D-02	0.84650D-02	0.88469D 00
30.	0.74820D-02	0.99528D-02	0.75359D 00
40.	0.74826D-02	0.12688D-01	0.59280D 00
45.	0.74829D-02	0.14860D-01	0.50726D 00
50.	0.74834D-02	0.17928D-01	0.42173D 00
60.	0.74845D-02	0.29295D-01	0.26102D 00
70.	0.74859D-02	0.60604D-01	0.13003D 00
80.	0.74877D-02	0.20022D 00	0.44551D-01
90.	0.74897D-02	0.10005D 01	0.14864D-01
100.	0.74920D-02	0.20039D 00	0.44546D-01
110.	0.74945D-02	0.60681D-01	0.13002D 00
120.	0.74970D-02	0.29346D-01	0.26101D 00
130.	0.74994D-02	0.17968D-01	0.42172D 00
135.	0.75006D-02	0.14895D-01	0.50725D 00
150.	0.75036D-02	0.99818D-02	0.75358D 00
165.	0.75056D-02	0.80406D-02	0.93396D 00
180.	0.75063D-02	0.75063D-02	0.10000D 01

ALPHA = 0.20

THETA	RHO S	RHO P	RHO U
0.	0.74755D-02	0.74755D-02	0.10000D 01
5.	0.74756D-02	0.75325D-02	0.99250D 00
10.	0.74758D-02	0.77070D-02	0.97021D 00
20.	0.74766D-02	0.84614D-02	0.88448D 00
30.	0.74781D-02	0.99527D-02	0.75321D 00
40.	0.74804D-02	0.12694D-01	0.59231D 00
45.	0.74819D-02	0.14872D-01	0.50676D 00
50.	0.74837D-02	0.17950D-01	0.42125D 00
60.	0.74881D-02	0.29351D-01	0.26066D 00
70.	0.74938D-02	0.60761D-01	0.12985D 00
80.	0.75008D-02	0.20079D 00	0.44522D-01
90.	0.75090D-02	0.10022D 01	0.14890D-01
100.	0.75182D-02	0.20147D 00	0.44500D-01
110.	0.75281D-02	0.61074D-01	0.12981D 00
120.	0.75383D-02	0.29558D-01	0.26061D 00
130.	0.75482D-02	0.18109D-01	0.42120D 00
135.	0.75529D-02	0.15016D-01	0.50671D 00
150.	0.75650D-02	0.10069D-01	0.75317D 00
165.	0.75731D-02	0.81141D-02	0.93383D 00
180.	0.75760D-02	0.75760D-02	0.10000D 01

ALPHA = 0.40

THETA	RHO S	RHO P	RHO U
0.	0.74523D-02	0.74523D-02	0.10000D 01
5.	0.74525D-02	0.75098D-02	0.99243D 00
10.	0.74533D-02	0.76858D-02	0.96998D 00
20.	0.74566D-02	0.84468D-02	0.88364D 00
30.	0.74625D-02	0.99521D-02	0.75169D 00
40.	0.74716D-02	0.12722D-01	0.59036D 00
45.	0.74777D-02	0.14924D-01	0.50476D 00
50.	0.74848D-02	0.18037D-01	0.41932D 00
60.	0.75026D-02	0.29578D-01	0.25921D 00
70.	0.75256D-02	0.61401D-01	0.12912D 00
80.	0.75539D-02	0.20314D 00	0.44405D-01
90.	0.75873D-02	0.10089D 01	0.14994D-01
100.	0.76249D-02	0.20592D 00	0.44315D-01
110.	0.76655D-02	0.62694D-01	0.12895D 00
120.	0.77073D-02	0.30432D-01	0.25897D 00
130.	0.77483D-02	0.18691D-01	0.41906D 00
135.	0.77677D-02	0.15515D-01	0.50449D 00
150.	0.78182D-02	0.10430D-01	0.75150D 00
165.	0.78520D-02	0.84181D-02	0.93327D 00
180.	0.78638D-02	0.78638D-02	0.10000D 01

ALPHA = 0.60

THETA	RHO S	RHO P	RHO U
0.	0.74139D-02	0.74139D-02	0.10000D 01
5.	0.74145D-02	0.74722D-02	0.99233D 00
10.	0.74162D-02	0.76507D-02	0.96958D 00
20.	0.74234D-02	0.84225D-02	0.88225D 00
30.	0.74366D-02	0.99512D-02	0.74917D 00
40.	0.74571D-02	0.12768D-01	0.58711D 00
45.	0.74706D-02	0.15011D-01	0.50141D 00
50.	0.74867D-02	0.18184D-01	0.41610D 00
60.	0.75271D-02	0.29967D-01	0.25677D 00
70.	0.75795D-02	0.62504D-01	0.12787D 00
80.	0.76446D-02	0.20718D 00	0.44204D-01
90.	0.77218D-02	0.10202D 01	0.15174D-01
100.	0.78095D-02	0.21379D 00	0.43994D-01
110.	0.79048D-02	0.65580D-01	0.12744D 00
120.	0.80037D-02	0.31995D-01	0.25611D 00
130.	0.81011D-02	0.19733D-01	0.41528D 00
135.	0.81475D-02	0.16410D-01	0.50058D 00
150.	0.82690D-02	0.11078D-01	0.74853D 00
165.	0.83504D-02	0.89625D-02	0.93227D 00
180.	0.83791D-02	0.83791D-02	0.10000D 01

ALPHA = 0.80

THETA	RHO S	RHO P	RHO U
0.	0.73610D-02	0.73610D-02	0.10000D 01
5.	0.73620D-02	0.74204D-02	0.99219D 00
10.	0.73650D-02	0.76021D-02	0.96904D 00
20.	0.73776D-02	0.83889D-02	0.88033D 00
30.	0.74005D-02	0.99499D-02	0.74566D 00
40.	0.74367D-02	0.12834D-01	0.58256D 00
45.	0.74607D-02	0.15134D-01	0.49672D 00
50.	0.74893D-02	0.18395D-01	0.41155D 00
60.	0.75618D-02	0.30533D-01	0.25330D 00
70.	0.76568D-02	0.64128D-01	0.12609D 00
80.	0.77759D-02	0.21316D 00	0.43914D-01
90.	0.79188D-02	0.10365D 01	0.15436D-01
100.	0.80829D-02	0.22586D 00	0.43519D-01
110.	0.82633D-02	0.70060D-01	0.12517D 00
120.	0.84523D-02	0.34436D-01	0.25177D 00
130.	0.86403D-02	0.21365D-01	0.40952D 00
135.	0.87305D-02	0.17812D-01	0.49457D 00
150.	0.89684D-02	0.12093D-01	0.74392D 00
165.	0.91294D-02	0.98157D-02	0.93071D 00
180.	0.91863D-02	0.91863D-02	0.10000D 01

ALPHA = 1.00

THETA	RHO S	RHO P	RHO U
0.	0.72943D-02	0.72943D-02	0.10000D 01
5.	0.72958D-02	0.73549D-02	0.99202D 00
10.	0.73003D-02	0.75407D-02	0.96836D 00
20.	0.73195D-02	0.83460D-02	0.87789D 00
30.	0.73547D-02	0.99482D-02	0.74120D 00
40.	0.74105D-02	0.12920D-01	0.57672D 00
45.	0.74480D-02	0.15298D-01	0.49066D 00
50.	0.74928D-02	0.18676D-01	0.40565D 00
60.	0.76074D-02	0.31301D-01	0.24876D 00
70.	0.77596D-02	0.66366D-01	0.12372D 00
80.	0.79530D-02	0.22143D 00	0.43523D-01
90.	0.81885D-02	0.10584D 01	0.15796D-01
100.	0.84631D-02	0.24348D 00	0.42859D-01
110.	0.87696D-02	0.76734D-01	0.12199D 00
120.	0.90956D-02	0.38105D-01	0.24556D 00
130.	0.94246D-02	0.23829D-01	0.40114D 00
135.	0.95840D-02	0.19931D-01	0.48579D 00
150.	0.10009D-01	0.13629D-01	0.73707D 00
165.	0.10301D-01	0.11104D-01	0.92837D 00
180.	0.10405D-01	0.10405D-01	0.10000D 01

ALPHA = 0.10

THETA	RHO S	RHO P	RHO U
0.	0.38142D-04	0.38142D-04	0.10000D 01
5.	0.38143D-04	0.38435D-04	0.99240D 00
10.	0.38144D-04	0.39331D-04	0.96983D 00
20.	0.38149D-04	0.43206D-04	0.88297D 00
30.	0.38158D-04	0.50884D-04	0.74990D 00
40.	0.38170D-04	0.65060D-04	0.58671D 00
45.	0.38177D-04	0.76374D-04	0.49989D 00
50.	0.38185D-04	0.92446D-04	0.41307D 00
60.	0.38203D-04	0.15286D-03	0.24995D 00
70.	0.38224D-04	0.32683D-03	0.11699D 00
80.	0.38248D-04	0.12674D-02	0.30215D-01
90.	0.38273D-04	0.10004D 01	0.76526D-04
100.	0.38299D-04	0.12696D-02	0.30202D-01
110.	0.38325D-04	0.32775D-03	0.11697D 00
120.	0.38350D-04	0.15347D-03	0.24992D 00
130.	0.38374D-04	0.92910D-04	0.41305D 00
135.	0.38385D-04	0.76794D-04	0.49986D 00
150.	0.38412D-04	0.51225D-04	0.74989D 00
165.	0.38430D-04	0.41191D-04	0.93297D 00
180.	0.38436D-04	0.38436D-04	0.10000D 01

ALPHA = 0.20

THETA	RHO S	RHO P	RHO U
0.	0.37927D-04	0.37927D-04	0.10000D 01
5.	0.37928D-04	0.38219D-04	0.99239D 00
10.	0.37934D-04	0.39116D-04	0.96978D 00
20.	0.37954D-04	0.42994D-04	0.88278D 00
30.	0.37988D-04	0.50682D-04	0.74956D 00
40.	0.38036D-04	0.64881D-04	0.58626D 00
45.	0.38065D-04	0.76220D-04	0.49943D 00
50.	0.38097D-04	0.92331D-04	0.41264D 00
60.	0.38170D-04	0.15293D-03	0.24962D 00
70.	0.38254D-04	0.32757D-03	0.11682D 00
80.	0.38348D-04	0.12724D-02	0.30176D-01
90.	0.38449D-04	0.10017D 01	0.76828D-04
100.	0.38554D-04	0.12814D-02	0.30125D-01
110.	0.38659D-04	0.33130D-03	0.11673D 00
120.	0.38762D-04	0.15538D-03	0.24950D 00
130.	0.38859D-04	0.94204D-04	0.41252D 00
135.	0.38903D-04	0.77915D-04	0.49932D 00
150.	0.39015D-04	0.52056D-04	0.74949D 00
165.	0.39087D-04	0.41901D-04	0.93284D 00
180.	0.39112D-04	0.39112D-04	0.10000D 01

ALPHA = 0.40

THETA	RHO S	RHO P	RHO U
0.	0.37066D-04	0.37066D-04	0.10000D 01
5.	0.37073D-04	0.37359D-04	0.99233D 00
10.	0.37093D-04	0.38258D-04	0.96956D 00
20.	0.37174D-04	0.42148D-04	0.88201D 00
30.	0.37310D-04	0.49869D-04	0.74817D 00
40.	0.37500D-04	0.64161D-04	0.58448D 00
45.	0.37614D-04	0.75594D-04	0.49760D 00
50.	0.37742D-04	0.91864D-04	0.41087D 00
60.	0.38036D-04	0.15321D-03	0.24828D 00
70.	0.38376D-04	0.33058D-03	0.11612D 00
80.	0.38758D-04	0.12928D-02	0.30016D-01
90.	0.39172D-04	0.10069D 01	0.78071D-04
100.	0.39607D-04	0.13305D-02	0.29806D-01
110.	0.40049D-04	0.34618D-03	0.11572D 00
120.	0.40483D-04	0.16342D-03	0.24775D 00
130.	0.40891D-04	0.99669D-04	0.41030D 00
135.	0.41080D-04	0.82653D-04	0.49704D 00
150.	0.41559D-04	0.55577D-04	0.74779D 00
165.	0.41871D-04	0.44913D-04	0.93228D 00
180.	0.41979D-04	0.41979D-04	0.10000D 01

ALPHA = 0.60

THETA	RHO S	RHO P	RHO U
0.	0.35641D-04	0.35641D-04	0.10000D 01
5.	0.35656D-04	0.35935D-04	0.99224D 00
10.	0.35701D-04	0.36835D-04	0.96920D 00
20.	0.35879D-04	0.40737D-04	0.88074D 00
30.	0.36177D-04	0.48505D-04	0.74585D 00
40.	0.36597D-04	0.62940D-04	0.58148D 00
45.	0.36854D-04	0.74529D-04	0.49451D 00
50.	0.37141D-04	0.91063D-04	0.40788D 00
60.	0.37805D-04	0.15370D-03	0.24600D 00
70.	0.38587D-04	0.33588D-03	0.11492D 00
80.	0.39474D-04	0.13291D-02	0.29738D-01
90.	0.40451D-04	0.10158D 01	0.80269D-04
100.	0.41493D-04	0.14213D-02	0.29235D-01
110.	0.42568D-04	0.37381D-03	0.11391D 00
120.	0.43636D-04	0.17846D-03	0.24455D 00
130.	0.44654D-04	0.10994D-03	0.40618D 00
135.	0.45129D-04	0.91581D-04	0.49280D 00
150.	0.46348D-04	0.62247D-04	0.74459D 00
165.	0.47149D-04	0.50632D-04	0.93121D 00
180.	0.47428D-04	0.47428D-04	0.10000D 01

ALPHA = 0.80

THETA	RHO S	RHO P	RHO U
0.	0.33666D-04	0.33666D-04	0.10000D 01
5.	0.33691D-04	0.33959D-04	0.99211D 00
10.	0.33767D-04	0.34858D-04	0.96870D 00
20.	0.34072D-04	0.38764D-04	0.87896D 00
30.	0.34586D-04	0.46575D-04	0.74260D 00
40.	0.35318D-04	0.61188D-04	0.57722D 00
45.	0.35769D-04	0.72986D-04	0.49010D 00
50.	0.36277D-04	0.89893D-04	0.40358D 00
60.	0.37470D-04	0.15443D-03	0.24267D 00
70.	0.38899D-04	0.34392D-03	0.11314D 00
80.	0.40554D-04	0.13851D-02	0.29318D-01
90.	0.42416D-04	0.10287D 01	0.83645D-04
100.	0.44447D-04	0.15708D-02	0.28340D-01
110.	0.46591D-04	0.41983D-03	0.11102D 00
120.	0.48770D-04	0.20380D-03	0.23934D 00
130.	0.50890D-04	0.12744D-03	0.39937D 00
135.	0.51895D-04	0.10685D-03	0.48571D 00
150.	0.54516D-04	0.73757D-04	0.73914D 00
165.	0.56271D-04	0.60549D-04	0.92935D 00
180.	0.56889D-04	0.56889D-04	0.10000D 01

ALPHA = 1.00

THETA	RHO S	RHO P	RHO U
0.	0.31159D-04	0.31159D-04	0.10000D 01
5.	0.31197D-04	0.31450D-04	0.99194D 00
10.	0.31309D-04	0.32342D-04	0.96806D 00
20.	0.31761D-04	0.36229D-04	0.87669D 00
30.	0.32532D-04	0.44059D-04	0.73840D 00
40.	0.33646D-04	0.58857D-04	0.57166D 00
45.	0.34339D-04	0.70910D-04	0.48428D 00
50.	0.35130D-04	0.88299D-04	0.39787D 00
60.	0.37016D-04	0.15544D-03	0.23816D 00
70.	0.39331D-04	0.35553D-03	0.11066D 00
80.	0.42093D-04	0.14677D-02	0.28720D-01
90.	0.45296D-04	0.10461D 01	0.88592D-04
100.	0.48908D-04	0.18153D-02	0.26989D-01
110.	0.52853D-04	0.49644D-03	0.10651D 00
120.	0.57007D-04	0.24685D-03	0.23098D 00
130.	0.61189D-04	0.15767D-03	0.38811D 00
135.	0.63220D-04	0.13343D-03	0.47385D 00
150.	0.68673D-04	0.94114D-04	0.72970D 00
165.	0.72450D-04	0.78235D-04	0.92607D 00
180.	0.73807D-04	0.73807D-04	0.10000D 01

ALPHA = 0.10

THETA	RHO S	RHO P	RHO U
0.	0.15195D-03	0.15195D-03	0.10000D 01
5.	0.15195D-03	0.15312D-03	0.99240D 00
10.	0.15196D-03	0.15668D-03	0.96984D 00
20.	0.15198D-03	0.17212D-03	0.88300D 00
30.	0.15201D-03	0.20270D-03	0.74997D 00
40.	0.15206D-03	0.25915D-03	0.58682D 00
45.	0.15208D-03	0.30420D-03	0.50002D 00
50.	0.15211D-03	0.36820D-03	0.41322D 00
60.	0.15218D-03	0.60869D-03	0.25013D 00
70.	0.15227D-03	0.13007D-02	0.11720D 00
80.	0.15236D-03	0.50291D-02	0.30443D-01
90.	0.15245D-03	0.10004D 01	0.30479D-03
100.	0.15255D-03	0.50398D-02	0.30418D-01
110.	0.15266D-03	0.13045D-02	0.11716D 00
120.	0.15275D-03	0.61111D-03	0.25008D 00
130.	0.15285D-03	0.37002D-03	0.41317D 00
135.	0.15289D-03	0.30585D-03	0.49996D 00
150.	0.15300D-03	0.20402D-03	0.74994D 00
165.	0.15307D-03	0.16406D-03	0.93299D 00
180.	0.15309D-03	0.15309D-03	0.10000D 01

ALPHA = 0.20

THETA	RHO S	RHO P	RHO U
0.	0.15123D-03	0.15123D-03	0.10000D 01
5.	0.15124D-03	0.15240D-03	0.99239D 00
10.	0.15126D-03	0.15597D-03	0.96979D 00
20.	0.15134D-03	0.17143D-03	0.88282D 00
30.	0.15147D-03	0.20206D-03	0.74965D 00
40.	0.15165D-03	0.25864D-03	0.58641D 00
45.	0.15176D-03	0.30381D-03	0.49960D 00
50.	0.15189D-03	0.36799D-03	0.41283D 00
60.	0.15217D-03	0.60932D-03	0.24985D 00
70.	0.15249D-03	0.13041D-02	0.11707D 00
80.	0.15286D-03	0.50489D-02	0.30424D-01
90.	0.15325D-03	0.10017D 01	0.30619D-03
100.	0.15365D-03	0.50921D-02	0.30324D-01
110.	0.15407D-03	0.13196D-02	0.11689D 00
120.	0.15447D-03	0.61908D-03	0.24962D 00
130.	0.15484D-03	0.37536D-03	0.41260D 00
135.	0.15501D-03	0.31045D-03	0.49939D 00
150.	0.15545D-03	0.20741D-03	0.74952D 00
165.	0.15573D-03	0.16694D-03	0.93285D 00
180.	0.15583D-03	0.15583D-03	0.10000D 01

ALPHA = 0.40

THETA	RHO S	RHO P	RHO U
0.	0.14837D-03	0.14837D-03	0.10000D 01
5.	0.14839D-03	0.14954D-03	0.99234D 00
10.	0.14847D-03	0.15313D-03	0.96959D 00
20.	0.14878D-03	0.16867D-03	0.88212D 00
30.	0.14931D-03	0.19951D-03	0.74838D 00
40.	0.15004D-03	0.25659D-03	0.58479D 00
45.	0.15048D-03	0.30224D-03	0.49796D 00
50.	0.15097D-03	0.36717D-03	0.41126D 00
60.	0.15210D-03	0.61187D-03	0.24870D 00
70.	0.15342D-03	0.13182D-02	0.11652D 00
80.	0.15490D-03	0.51296D-02	0.30347D-01
90.	0.15650D-03	0.10069D 01	0.31188D-03
100.	0.15819D-03	0.53111D-02	0.29938D-01
110.	0.15990D-03	0.13829D-02	0.11577D 00
120.	0.16159D-03	0.65258D-03	0.24774D 00
130.	0.16317D-03	0.39784D-03	0.41025D 00
135.	0.16391D-03	0.32986D-03	0.49698D 00
150.	0.16577D-03	0.22171D-03	0.74775D 00
165.	0.16698D-03	0.17912D-03	0.93226D 00
180.	0.16740D-03	0.16740D-03	0.10000D 01

ALPHA = 0.60

THETA	RHO S	RHO P	RHO U
0.	0.14366D-03	0.14366D-03	0.10000D 01
5.	0.14371D-03	0.14484D-03	0.99225D 00
10.	0.14388D-03	0.14845D-03	0.96926D 00
20.	0.14457D-03	0.16411D-03	0.88095D 00
30.	0.14572D-03	0.19527D-03	0.74627D 00
40.	0.14734D-03	0.25314D-03	0.58210D 00
45.	0.14832D-03	0.29957D-03	0.49520D 00
50.	0.14943D-03	0.36578D-03	0.40862D 00
60.	0.15200D-03	0.61625D-03	0.24676D 00
70.	0.15501D-03	0.13427D-02	0.11558D 00
80.	0.15844D-03	0.52709D-02	0.30213D-01
90.	0.16222D-03	0.10156D 01	0.32189D-03
100.	0.16625D-03	0.57150D-02	0.29251D-01
110.	0.17041D-03	0.15000D-02	0.11376D 00
120.	0.17454D-03	0.71485D-03	0.24430D 00
130.	0.17849D-03	0.43983D-03	0.40591D 00
135.	0.18033D-03	0.36618D-03	0.49254D 00
150.	0.18505D-03	0.24859D-03	0.74444D 00
165.	0.18815D-03	0.20207D-03	0.93116D 00
180.	0.18924D-03	0.18924D-03	0.10000D 01

ALPHA = 0.80

THETA	RHO S	RHO P	RHO U
0.	0.13719D-03	0.13719D-03	0.10000D 01
5.	0.13729D-03	0.13838D-03	0.99213D 00
10.	0.13758D-03	0.14201D-03	0.96880D 00
20.	0.13875D-03	0.15779D-03	0.87934D 00
30.	0.14073D-03	0.18933D-03	0.74334D 00
40.	0.14355D-03	0.24825D-03	0.57831D 00
45.	0.14528D-03	0.29576D-03	0.49130D 00
50.	0.14724D-03	0.36376D-03	0.40487D 00
60.	0.15184D-03	0.62269D-03	0.24396D 00
70.	0.15734D-03	0.13794D-02	0.11421D 00
80.	0.16372D-03	0.54838D-02	0.30014D-01
90.	0.17090D-03	0.10279D 01	0.33709D-03
100.	0.17872D-03	0.63799D-02	0.28187D-01
110.	0.18698D-03	0.16935D-02	0.11058D 00
120.	0.19538D-03	0.81876D-03	0.23877D 00
130.	0.20354D-03	0.51052D-03	0.39882D 00
135.	0.20741D-03	0.42756D-03	0.48521D 00
150.	0.21750D-03	0.29440D-03	0.73887D 00
165.	0.22426D-03	0.24133D-03	0.92928D 00
180.	0.22664D-03	0.22664D-03	0.10000D 01

ALPHA = 1.00

THETA	RHO S	RHO P	RHO U
0.	0.12909D-03	0.12909D-03	0.10000D 01
5.	0.12923D-03	0.13027D-03	0.99199D 00
10.	0.12966D-03	0.13392D-03	0.96823D 00
20.	0.13141D-03	0.14979D-03	0.87731D 00
30.	0.13438D-03	0.18169D-03	0.73961D 00
40.	0.13866D-03	0.24184D-03	0.57342D 00
45.	0.14133D-03	0.29071D-03	0.48624D 00
50.	0.14437D-03	0.36106D-03	0.39995D 00
60.	0.15163D-03	0.63154D-03	0.24021D 00
70.	0.16053D-03	0.14309D-02	0.11233D 00
80.	0.17112D-03	0.57873D-02	0.29735D-01
90.	0.18340D-03	0.10441D 01	0.35899D-03
100.	0.19723D-03	0.74656D-02	0.26610D-01
110.	0.21231D-03	0.20117D-02	0.10573D 00
120.	0.22815D-03	0.99235D-03	0.23008D 00
130.	0.24407D-03	0.63036D-03	0.38735D 00
135.	0.25180D-03	0.53225D-03	0.47321D 00
150.	0.27249D-03	0.37361D-03	0.72943D 00
165.	0.28681D-03	0.30972D-03	0.92602D 00
180.	0.29194D-03	0.29194D-03	0.10000D 01

ALPHA = 0.10

THETA	RHO S	RHO P	RHO U
0.	0.59891D-03	0.59891D-03	0.10000D 01
5.	0.59892D-03	0.60350D-03	0.99241D 00
10.	0.59894D-03	0.61756D-03	0.96987D 00
20.	0.59901D-03	0.67835D-03	0.88311D 00
30.	0.59914D-03	0.79878D-03	0.75021D 00
40.	0.59931D-03	0.10210D-02	0.58721D 00
45.	0.59941D-03	0.11984D-02	0.50049D 00
50.	0.59953D-03	0.14501D-02	0.41378D 00
60.	0.59979D-03	0.23955D-02	0.25083D 00
70.	0.60009D-03	0.51078D-02	0.11802D 00
80.	0.60043D-03	0.19531D-01	0.31324D-01
90.	0.60080D-03	0.10004D 01	0.12006D-02
100.	0.60118D-03	0.19587D-01	0.31275D-01
110.	0.60156D-03	0.51241D-02	0.11793D 00
120.	0.60193D-03	0.24051D-02	0.25073D 00
130.	0.60228D-03	0.14572D-02	0.41367D 00
135.	0.60244D-03	0.12047D-02	0.50039D 00
150.	0.60285D-03	0.80380D-03	0.75015D 00
165.	0.60311D-03	0.64642D-03	0.93304D 00
180.	0.60320D-03	0.60320D-03	0.10000D 01

ALPHA = 0.20

THETA	RHO S	RHO P	RHO U
0.	0.59710D-03	0.59710D-03	0.10000D 01
5.	0.59713D-03	0.60170D-03	0.99240D 00
10.	0.59720D-03	0.61579D-03	0.96983D 00
20.	0.59750D-03	0.67675D-03	0.88296D 00
30.	0.59799D-03	0.79753D-03	0.74995D 00
40.	0.59867D-03	0.10205D-02	0.58689D 00
45.	0.59909D-03	0.11985D-02	0.50017D 00
50.	0.59955D-03	0.14512D-02	0.41348D 00
60.	0.60060D-03	0.24006D-02	0.25064D 00
70.	0.60182D-03	0.51253D-02	0.11795D 00
80.	0.60318D-03	0.19608D-01	0.31346D-01
90.	0.60465D-03	0.10017D 01	0.12075D-02
100.	0.60618D-03	0.19832D-01	0.31153D-01
110.	0.60773D-03	0.51912D-02	0.11761D 00
120.	0.60924D-03	0.24394D-02	0.25021D 00
130.	0.61066D-03	0.14797D-02	0.41304D 00
135.	0.61131D-03	0.12240D-02	0.49976D 00
150.	0.61296D-03	0.81778D-03	0.74969D 00
165.	0.61403D-03	0.65823D-03	0.93290D 00
180.	0.61440D-03	0.61440D-03	0.10000D 01

ALPHA = 0.40

THETA	RHO S	RHO P	RHO U
0.	0.58995D-03	0.58995D-03	0.10000D 01
5.	0.59004D-03	0.59459D-03	0.99235D 00
10.	0.59033D-03	0.60882D-03	0.96966D 00
20.	0.59149D-03	0.67040D-03	0.88237D 00
30.	0.59343D-03	0.79256D-03	0.74891D 00
40.	0.59615D-03	0.10184D-02	0.58560D 00
45.	0.59780D-03	0.11990D-02	0.49889D 00
50.	0.59963D-03	0.14556D-02	0.41230D 00
60.	0.60386D-03	0.24211D-02	0.24987D 00
70.	0.60878D-03	0.51959D-02	0.11770D 00
80.	0.61431D-03	0.19916D-01	0.31440D-01
90.	0.62033D-03	0.10068D 01	0.12357D-02
100.	0.62667D-03	0.20857D-01	0.30653D-01
110.	0.63313D-03	0.54714D-02	0.11628D 00
120.	0.63948D-03	0.25829D-02	0.24806D 00
130.	0.64545D-03	0.15741D-02	0.41044D 00
135.	0.64822D-03	0.13048D-02	0.49712D 00
150.	0.65527D-03	0.87647D-03	0.74779D 00
165.	0.65985D-03	0.70782D-03	0.93227D 00
180.	0.66144D-03	0.66144D-03	0.10000D 01

ALPHA = 0.60

THETA	RHO S	RHO P	RHO U
0.	0.57832D-03	0.57832D-03	0.10000D 01
5.	0.57853D-03	0.58304D-03	0.99228D 00
10.	0.57917D-03	0.59747D-03	0.96938D 00
20.	0.58171D-03	0.66002D-03	0.88141D 00
30.	0.58597D-03	0.78438D-03	0.74720D 00
40.	0.59199D-03	0.10150D-02	0.58347D 00
45.	0.59566D-03	0.11998D-02	0.49676D 00
50.	0.59977D-03	0.14630D-02	0.41032D 00
60.	0.60932D-03	0.24558D-02	0.24858D 00
70.	0.62057D-03	0.53163D-02	0.11728D 00
80.	0.63338D-03	0.20441D-01	0.31599D-01
90.	0.64750D-03	0.10151D 01	0.12845D-02
100.	0.66259D-03	0.22743D-01	0.29777D-01
110.	0.67818D-03	0.59850D-02	0.11391D 00
120.	0.69370D-03	0.28467D-02	0.24421D 00
130.	0.70850D-03	0.17481D-02	0.40571D 00
135.	0.71541D-03	0.14542D-02	0.49232D 00
150.	0.73317D-03	0.98532D-03	0.74428D 00
165.	0.74483D-03	0.79999D-03	0.93111D 00
180.	0.74891D-03	0.74891D-03	0.10000D 01

ALPHA = 0.80

THETA	RHO S	RHO P	RHO U
0.	0.56266D-03	0.56266D-03	0.10000D 01
5.	0.56302D-03	0.56746D-03	0.99219D 00
10.	0.56410D-03	0.58215D-03	0.96901D 00
20.	0.56845D-03	0.64593D-03	0.88011D 00
30.	0.57579D-03	0.77316D-03	0.74487D 00
40.	0.58625D-03	0.10103D-02	0.58055D 00
45.	0.59270D-03	0.12010D-02	0.49382D 00
50.	0.59997D-03	0.14733D-02	0.40758D 00
60.	0.61704D-03	0.25054D-02	0.24675D 00
70.	0.63749D-03	0.54905D-02	0.11667D 00
80.	0.66121D-03	0.21198D-01	0.31832D-01
90.	0.68789D-03	0.10266D 01	0.13570D-02
100.	0.71699D-03	0.25835D-01	0.28449D-01
110.	0.74770D-03	0.68216D-02	0.11027D 00
120.	0.77891D-03	0.32786D-02	0.23817D 00
130.	0.80925D-03	0.20350D-02	0.39816D 00
135.	0.82362D-03	0.17011D-02	0.48460D 00
150.	0.86110D-03	0.11663D-02	0.73852D 00
165.	0.88617D-03	0.95377D-03	0.92919D 00
180.	0.89500D-03	0.89500D-03	0.10000D 01

ALPHA = 1.00

THETA	RHO S	RHO P	RHO U
0.	0.54351D-03	0.54351D-03	0.10000D 01
5.	0.54404D-03	0.54839D-03	0.99207D 00
10.	0.54564D-03	0.56337D-03	0.96856D 00
20.	0.55212D-03	0.62852D-03	0.87851D 00
30.	0.56315D-03	0.75913D-03	0.74197D 00
40.	0.57905D-03	0.10042D-02	0.57687D 00
45.	0.58895D-03	0.12024D-02	0.49009D 00
50.	0.60022D-03	0.14868D-02	0.40407D 00
60.	0.62708D-03	0.25711D-02	0.24437D 00
70.	0.65997D-03	0.57249D-02	0.11586D 00
80.	0.69908D-03	0.22213D-01	0.32148D-01
90.	0.74429D-03	0.10408D 01	0.14583D-02
100.	0.79507D-03	0.30853D-01	0.26544D-01
110.	0.85029D-03	0.81650D-02	0.10490D 00
120.	0.90815D-03	0.39780D-02	0.22900D 00
130.	0.96613D-03	0.25042D-02	0.38640D 00
135.	0.99418D-03	0.21068D-02	0.47242D 00
150.	0.10692D-02	0.14669D-02	0.72915D 00
165.	0.11209D-02	0.12105D-02	0.92599D 00
180.	0.11393D-02	0.11393D-02	0.10000D 01

ALPHA = 0.10

THETA	RHO S	RHO P	RHO U
0.	0.13177D-02	0.13177D-02	0.10000D 01
5.	0.13177D-02	0.13278D-02	0.99242D 00
10.	0.13177D-02	0.13587D-02	0.96991D 00
20.	0.13179D-02	0.14923D-02	0.88329D 00
30.	0.13182D-02	0.17569D-02	0.75059D 00
40.	0.13185D-02	0.22451D-02	0.58783D 00
45.	0.13187D-02	0.26344D-02	0.50124D 00
50.	0.13190D-02	0.31868D-02	0.41465D 00
60.	0.13195D-02	0.52580D-02	0.25194D 00
70.	0.13202D-02	0.11174D-01	0.11931D 00
80.	0.13209D-02	0.41996D-01	0.32730D-01
90.	0.13217D-02	0.10004D 01	0.26393D-02
100.	0.13225D-02	0.42143D-01	0.32660D-01
110.	0.13233D-02	0.11213D-01	0.11918D 00
120.	0.13241D-02	0.52795D-02	0.25178D 00
130.	0.13248D-02	0.32022D-02	0.41449D 00
135.	0.13252D-02	0.26480D-02	0.50109D 00
150.	0.13260D-02	0.17676D-02	0.75050D 00
165.	0.13266D-02	0.14218D-02	0.93314D 00
180.	0.13268D-02	0.13268D-02	0.10000D 01

ALPHA = 0.20

THETA	RHO S	RHO P	RHO U
0.	0.13158D-02	0.13158D-02	0.10000D 01
5.	0.13158D-02	0.13259D-02	0.99241D 00
10.	0.13160D-02	0.13569D-02	0.96988D 00
20.	0.13166D-02	0.14910D-02	0.88317D 00
30.	0.13176D-02	0.17567D-02	0.75038D 00
40.	0.13191D-02	0.22469D-02	0.58759D 00
45.	0.13199D-02	0.26380D-02	0.50100D 00
50.	0.13209D-02	0.31931D-02	0.41445D 00
60.	0.13231D-02	0.52745D-02	0.25184D 00
70.	0.13257D-02	0.11220D-01	0.11932D 00
80.	0.13286D-02	0.42157D-01	0.32799D-01
90.	0.13317D-02	0.10017D 01	0.26575D-02
100.	0.13349D-02	0.42751D-01	0.32517D-01
110.	0.13382D-02	0.11376D-01	0.11881D 00
120.	0.13414D-02	0.53613D-02	0.25121D 00
130.	0.13444D-02	0.32551D-02	0.41381D 00
135.	0.13458D-02	0.26931D-02	0.50040D 00
150.	0.13493D-02	0.17999D-02	0.75001D 00
165.	0.13516D-02	0.14488D-02	0.93298D 00
180.	0.13524D-02	0.13524D-02	0.10000D 01

ALPHA = 0.40

THETA	RHO S	RHO P	RHO U
0.	0.13082D-02	0.13082D-02	0.10000D 01
5.	0.13084D-02	0.13184D-02	0.99238D 00
10.	0.13090D-02	0.13499D-02	0.96974D 00
20.	0.13114D-02	0.14860D-02	0.88269D 00
30.	0.13155D-02	0.17558D-02	0.74956D 00
40.	0.13212D-02	0.22543D-02	0.58662D 00
45.	0.13246D-02	0.26524D-02	0.50008D 00
50.	0.13285D-02	0.32178D-02	0.41364D 00
60.	0.13374D-02	0.53402D-02	0.25144D 00
70.	0.13478D-02	0.11404D-01	0.11937D 00
80.	0.13594D-02	0.42800D-01	0.33077D-01
90.	0.13722D-02	0.10067D 01	0.27314D-02
100.	0.13856D-02	0.45292D-01	0.31933D-01
110.	0.13993D-02	0.12056D-01	0.11730D 00
120.	0.14127D-02	0.57016D-02	0.24884D 00
130.	0.14254D-02	0.34755D-02	0.41098D 00
135.	0.14313D-02	0.28808D-02	0.49756D 00
150.	0.14463D-02	0.19346D-02	0.74799D 00
165.	0.14561D-02	0.15619D-02	0.93232D 00
180.	0.14594D-02	0.14594D-02	0.10000D 01

ALPHA = 0.60

THETA	RHO S	RHO P	RHO U
0.	0.12960D-02	0.12960D-02	0.10000D 01
5.	0.12965D-02	0.13065D-02	0.99232D 00
10.	0.12978D-02	0.13386D-02	0.96952D 00
20.	0.13031D-02	0.14778D-02	0.88193D 00
30.	0.13120D-02	0.17543D-02	0.74823D 00
40.	0.13246D-02	0.22663D-02	0.58504D 00
45.	0.13323D-02	0.26759D-02	0.49856D 00
50.	0.13410D-02	0.32586D-02	0.41231D 00
60.	0.13610D-02	0.54494D-02	0.25078D 00
70.	0.13847D-02	0.11711D-01	0.11946D 00
80.	0.14117D-02	0.43865D-01	0.33546D-01
90.	0.14414D-02	0.10147D 01	0.28578D-02
100.	0.14733D-02	0.49948D-01	0.30924D-01
110.	0.15062D-02	0.13290D-01	0.11466D 00
120.	0.15390D-02	0.63208D-02	0.24465D 00
130.	0.15703D-02	0.38774D-02	0.40593D 00
135.	0.15849D-02	0.32237D-02	0.49247D 00
150.	0.16225D-02	0.21811D-02	0.74431D 00
165.	0.16472D-02	0.17693D-02	0.93111D 00
180.	0.16559D-02	0.16559D-02	0.10000D 01

ALPHA = 0.80

THETA	RHO S	RHO P	RHO U
0.	0.12799D-02	0.12799D-02	0.10000D 01
5.	0.12807D-02	0.12907D-02	0.99224D 00
10.	0.12829D-02	0.13237D-02	0.96923D 00
20.	0.12920D-02	0.14669D-02	0.88091D 00
30.	0.13074D-02	0.17522D-02	0.74646D 00
40.	0.13293D-02	0.22826D-02	0.58291D 00
45.	0.13428D-02	0.27082D-02	0.49650D 00
50.	0.13580D-02	0.33147D-02	0.41050D 00
60.	0.13938D-02	0.56018D-02	0.24986D 00
70.	0.14367D-02	0.12142D-01	0.11958D 00
80.	0.14864D-02	0.45342D-01	0.34217D-01
90.	0.15424D-02	0.10253D 01	0.30419D-02
100.	0.16034D-02	0.57530D-01	0.29427D-01
110.	0.16678D-02	0.15274D-01	0.11068D 00
120.	0.17332D-02	0.73169D-02	0.23820D 00
130.	0.17968D-02	0.45267D-02	0.39802D 00
135.	0.18270D-02	0.37787D-02	0.48443D 00
150.	0.19055D-02	0.25823D-02	0.73841D 00
165.	0.19580D-02	0.21076D-02	0.92915D 00
180.	0.19765D-02	0.19765D-02	0.10000D 01

ALPHA = 1.00

THETA	RHO S	RHO P	RHO U
0.	0.12607D-02	0.12607D-02	0.10000D 01
5.	0.12618D-02	0.12718D-02	0.99215D 00
10.	0.12651D-02	0.13058D-02	0.96887D 00
20.	0.12787D-02	0.14538D-02	0.87969D 00
30.	0.13018D-02	0.17498D-02	0.74430D 00
40.	0.13350D-02	0.23028D-02	0.58030D 00
45.	0.13557D-02	0.27485D-02	0.49395D 00
50.	0.13793D-02	0.33854D-02	0.40823D 00
60.	0.14354D-02	0.57967D-02	0.24870D 00
70.	0.15039D-02	0.12700D-01	0.11974D 00
80.	0.15854D-02	0.47216D-01	0.35106D-01
90.	0.16793D-02	0.10380D 01	0.32917D-02
100.	0.17846D-02	0.69695D-01	0.27342D-01
110.	0.18989D-02	0.18387D-01	0.10497D 00
120.	0.20184D-02	0.88845D-02	0.22874D 00
130.	0.21378D-02	0.55554D-02	0.38613D 00
135.	0.21955D-02	0.46609D-02	0.47221D 00
150.	0.23494D-02	0.32249D-02	0.72915D 00
165.	0.24552D-02	0.26519D-02	0.92602D 00
180.	0.24930D-02	0.24930D-02	0.10000D 01

ALPHA = 0.10

THETA	RHO S	RHO P	RHO U
0.	0.22758D-02	0.22758D-02	0.10000D 01
5.	0.22758D-02	0.22932D-02	0.99244D 00
10.	0.22759D-02	0.23465D-02	0.96997D 00
20.	0.22761D-02	0.25770D-02	0.88352D 00
30.	0.22766D-02	0.30333D-02	0.75108D 00
40.	0.22772D-02	0.38747D-02	0.58865D 00
45.	0.22775D-02	0.45452D-02	0.50222D 00
50.	0.22779D-02	0.54960D-02	0.41580D 00
60.	0.22789D-02	0.90538D-02	0.25340D 00
70.	0.22799D-02	0.19157D-01	0.12102D 00
80.	0.22811D-02	0.70414D-01	0.34598D-01
90.	0.22824D-02	0.10004D 01	0.45535D-02
100.	0.22838D-02	0.70703D-01	0.34506D-01
110.	0.22852D-02	0.19228D-01	0.12085D 00
120.	0.22865D-02	0.90917D-02	0.25320D 00
130.	0.22877D-02	0.55225D-02	0.41560D 00
135.	0.22883D-02	0.45685D-02	0.50203D 00
150.	0.22898D-02	0.30514D-02	0.75097D 00
165.	0.22907D-02	0.24549D-02	0.93326D 00
180.	0.22910D-02	0.22910D-02	0.10000D 01

ALPHA = 0.20

THETA	RHO S	RHO P	RHO U
0.	0.22757D-02	0.22757D-02	0.10000D 01
5.	0.22758D-02	0.22932D-02	0.99243D 00
10.	0.22760D-02	0.23467D-02	0.96995D 00
20.	0.22770D-02	0.25783D-02	0.88343D 00
30.	0.22788D-02	0.30369D-02	0.75093D 00
40.	0.22812D-02	0.38825D-02	0.58848D 00
45.	0.22826D-02	0.45566D-02	0.50208D 00
50.	0.22842D-02	0.55126D-02	0.41570D 00
60.	0.22879D-02	0.90902D-02	0.25340D 00
70.	0.22922D-02	0.19247D-01	0.12111D 00
80.	0.22970D-02	0.70674D-01	0.34719D-01
90.	0.23022D-02	0.10017D 01	0.45900D-02
100.	0.23077D-02	0.71844D-01	0.34349D-01
110.	0.23132D-02	0.19535D-01	0.12045D 00
120.	0.23186D-02	0.92435D-02	0.25257D 00
130.	0.23237D-02	0.56195D-02	0.41486D 00
135.	0.23260D-02	0.46508D-02	0.50129D 00
150.	0.23320D-02	0.31098D-02	0.75045D 00
165.	0.23358D-02	0.25037D-02	0.93310D 00
180.	0.23371D-02	0.23371D-02	0.10000D 01

ALPHA = 0.40

THETA	RHO S	RHO P	RHO U
0.	0.22752D-02	0.22752D-02	0.10000D 01
5.	0.22755D-02	0.22930D-02	0.99240D 00
10.	0.22765D-02	0.23475D-02	0.96984D 00
20.	0.22806D-02	0.25833D-02	0.88307D 00
30.	0.22873D-02	0.30507D-02	0.75033D 00
40.	0.22968D-02	0.39136D-02	0.58783D 00
45.	0.23026D-02	0.46018D-02	0.50151D 00
50.	0.23090D-02	0.55785D-02	0.41527D 00
60.	0.23239D-02	0.92347D-02	0.25338D 00
70.	0.23412D-02	0.19604D-01	0.12148D 00
80.	0.23607D-02	0.71700D-01	0.35202D-01
90.	0.23820D-02	0.10066D 01	0.47370D-02
100.	0.24045D-02	0.76598D-01	0.33714D-01
110.	0.24274D-02	0.20809D-01	0.11879D 00
120.	0.24501D-02	0.98723D-02	0.25001D 00
130.	0.24714D-02	0.60221D-02	0.41184D 00
135.	0.24813D-02	0.49922D-02	0.49827D 00
150.	0.25065D-02	0.33523D-02	0.74832D 00
165.	0.25229D-02	0.27063D-02	0.93241D 00
180.	0.25286D-02	0.25286D-02	0.10000D 01

ALPHA = 0.60

THETA	RHO S	RHO P	RHO U
0.	0.22744D-02	0.22744D-02	0.10000D 01
5.	0.22752D-02	0.22927D-02	0.99236D 00
10.	0.22774D-02	0.23488D-02	0.96967D 00
20.	0.22862D-02	0.25914D-02	0.88249D 00
30.	0.23010D-02	0.30730D-02	0.74938D 00
40.	0.23220D-02	0.39636D-02	0.58679D 00
45.	0.23348D-02	0.46750D-02	0.50059D 00
50.	0.23492D-02	0.56854D-02	0.41457D 00
60.	0.23826D-02	0.94711D-02	0.25334D 00
70.	0.24220D-02	0.20190D-01	0.12209D 00
80.	0.24670D-02	0.73356D-01	0.36009D-01
90.	0.25167D-02	0.10144D 01	0.49852D-02
100.	0.25699D-02	0.85258D-01	0.32629D-01
110.	0.26250D-02	0.23104D-01	0.11594D 00
120.	0.26799D-02	0.11005D-01	0.24554D 00
130.	0.27323D-02	0.67484D-02	0.40650D 00
135.	0.27568D-02	0.56087D-02	0.49292D 00
150.	0.28198D-02	0.37912D-02	0.74449D 00
165.	0.28612D-02	0.30734D-02	0.93116D 00
180.	0.28756D-02	0.28756D-02	0.10000D 01

ALPHA = 0.80

THETA	RHO S	RHO P	RHO U
0.	0.22735D-02	0.22735D-02	0.10000D 01
5.	0.22747D-02	0.22924D-02	0.99230D 00
10.	0.22785D-02	0.23504D-02	0.96945D 00
20.	0.22935D-02	0.26020D-02	0.88174D 00
30.	0.23191D-02	0.31023D-02	0.74812D 00
40.	0.23554D-02	0.40302D-02	0.58541D 00
45.	0.23778D-02	0.47732D-02	0.49936D 00
50.	0.24031D-02	0.58296D-02	0.41364D 00
60.	0.24625D-02	0.97930D-02	0.25330D 00
70.	0.25337D-02	0.20991D-01	0.12293D 00
80.	0.26164D-02	0.75569D-01	0.37141D-01
90.	0.27093D-02	0.10243D 01	0.53400D-02
100.	0.28108D-02	0.99227D-01	0.31050D-01
110.	0.29178D-02	0.26741D-01	0.11171D 00
120.	0.30265D-02	0.12798D-01	0.23879D 00
130.	0.31322D-02	0.79006D-02	0.39833D 00
135.	0.31822D-02	0.65882D-02	0.48465D 00
150.	0.33126D-02	0.44910D-02	0.73848D 00
165.	0.33998D-02	0.36599D-02	0.92917D 00
180.	0.34305D-02	0.34305D-02	0.10000D 01

ALPHA = 1.00

THETA	RHO S	RHO P	RHO U
0.	0.22723D-02	0.22723D-02	0.10000D 01
5.	0.22742D-02	0.22920D-02	0.99223D 00
10.	0.22797D-02	0.23524D-02	0.96919D 00
20.	0.23022D-02	0.26144D-02	0.88086D 00
30.	0.23405D-02	0.31372D-02	0.74662D 00
40.	0.23956D-02	0.41107D-02	0.58375D 00
45.	0.24298D-02	0.48924D-02	0.49787D 00
50.	0.24688D-02	0.60062D-02	0.41250D 00
60.	0.25616D-02	0.10192D-01	0.25324D 00
70.	0.26749D-02	0.21989D-01	0.12399D 00
80.	0.28092D-02	0.78245D-01	0.38604D-01
90.	0.29641D-02	0.10356D 01	0.58090D-02
100.	0.31373D-02	0.12130D 00	0.28911D-01
110.	0.33250D-02	0.32328D-01	0.10582D 00
120.	0.35208D-02	0.15547D-01	0.22917D 00
130.	0.37162D-02	0.96745D-02	0.38640D 00
135.	0.38105D-02	0.80996D-02	0.47246D 00
150.	0.40615D-02	0.55770D-02	0.72936D 00
165.	0.42338D-02	0.45732D-02	0.92610D 00
180.	0.42953D-02	0.42953D-02	0.10000D 01

ALPHA = 0.02

THETA	RHO S	RHO P	RHO U
0.	0.62875D-04	0.62875D-04	0.10000D 01
5.	0.62842D-04	0.63322D-04	0.99242D 00
10.	0.62745D-04	0.64691D-04	0.96991D 00
20.	0.62360D-04	0.70601D-04	0.88327D 00
30.	0.61730D-04	0.82255D-04	0.75049D 00
40.	0.60877D-04	0.10362D-03	0.58755D 00
45.	0.60374D-04	0.12056D-03	0.50081D 00
50.	0.59825D-04	0.14450D-03	0.41404D 00
60.	0.58606D-04	0.23365D-03	0.25087D 00
70.	0.57258D-04	0.48667D-03	0.11770D 00
80.	0.55823D-04	0.18270D-02	0.30608D-01
90.	0.54342D-04	0.99993D 00	0.10868D-03
100.	0.52863D-04	0.17700D-02	0.29917D-01
110.	0.51429D-04	0.44175D-03	0.11647D 00
120.	0.50084D-04	0.20090D-03	0.24933D 00
130.	0.48869D-04	0.11848D-03	0.41249D 00
135.	0.48321D-04	0.96772D-04	0.49936D 00
150.	0.46969D-04	0.62660D-04	0.74960D 00
165.	0.46120D-04	0.49438D-04	0.93289D 00
180.	0.45830D-04	0.45830D-04	0.10000D 01

ALPHA = 0.04

THETA	RHO S	RHO P	RHO U
0.	0.88723D-04	0.88723D-04	0.10000D 01
5.	0.88592D-04	0.89264D-04	0.99247D 00
10.	0.88198D-04	0.90916D-04	0.97011D 00
20.	0.86640D-04	0.98012D-04	0.88399D 00
30.	0.84099D-04	0.11185D-03	0.75190D 00
40.	0.80652D-04	0.13680D-03	0.58959D 00
45.	0.78623D-04	0.15629D-03	0.50308D 00
50.	0.76409D-04	0.18349D-03	0.41646D 00
60.	0.71502D-04	0.28239D-03	0.25325D 00
70.	0.66082D-04	0.55277D-03	0.11961D 00
80.	0.60317D-04	0.19088D-02	0.31657D-01
90.	0.54384D-04	0.99973D 00	0.10878D-03
100.	0.48463D-04	0.16803D-02	0.28889D-01
110.	0.42734D-04	0.37288D-03	0.11464D 00
120.	0.37368D-04	0.15125D-03	0.24709D 00
130.	0.32528D-04	0.79290D-04	0.41026D 00
135.	0.30350D-04	0.61035D-04	0.49727D 00
150.	0.24976D-04	0.33376D-04	0.74834D 00
165.	0.21604D-04	0.23167D-04	0.93252D 00
180.	0.20454D-04	0.20454D-04	0.10000D 01

ALPHA = 0.06

THETA	RHO S	RHO P	RHO U
0.	0.13252D-03	0.13252D-03	0.10000D 01
5.	0.13222D-03	0.13321D-03	0.99255D 00
10.	0.13131D-03	0.13532D-03	0.97044D 00
20.	0.12775D-03	0.14432D-03	0.88520D 00
30.	0.12193D-03	0.16166D-03	0.75428D 00
40.	0.11405D-03	0.19234D-03	0.59304D 00
45.	0.10942D-03	0.21588D-03	0.50692D 00
50.	0.10438D-03	0.24824D-03	0.42053D 00
60.	0.93209D-04	0.36240D-03	0.25727D 00
70.	0.80905D-04	0.65922D-03	0.12280D 00
80.	0.67851D-04	0.20345D-02	0.33417D-01
90.	0.54454D-04	0.99938D 00	0.10893D-03
100.	0.41121D-04	0.15154D-02	0.27175D-01
110.	0.28256D-04	0.25323D-03	0.11161D 00
120.	0.16239D-04	0.66732D-04	0.24336D 00
130.	0.54241D-05	0.13342D-04	0.40655D 00
135.	0.56634D-06	0.11469D-05	0.49381D 00
150.	0.56634D-06	0.11469D-05	
165.	0.56634D-06	0.11469D-05	
180.	0.56634D-06	0.11469D-05	

ALPHA = 0.08

THETA	RHO S	RHO P	RHO U
0.	0.19537D-03	0.19537D-03	0.10000D 01
5.	0.19482D-03	0.19626D-03	0.99267D 00
10.	0.19318D-03	0.19897D-03	0.97091D 00
20.	0.18667D-03	0.21047D-03	0.88694D 00
30.	0.17608D-03	0.23241D-03	0.75768D 00
40.	0.16177D-03	0.27057D-03	0.59797D 00
45.	0.15337D-03	0.29937D-03	0.51240D 00
50.	0.14423D-03	0.33837D-03	0.42634D 00
60.	0.12406D-03	0.47192D-03	0.26296D 00
70.	0.10190D-03	0.80091D-03	0.12732D 00
80.	0.78489D-04	0.21909D-02	0.35900D-01
90.	0.54551D-04	0.99890D 00	0.10916D-03
100.	0.30824D-04	0.12459D-02	0.24770D-01
110.	0.80162D-05	0.74671D-04	0.10736D 00
120.	0.80162D-05	0.74671D-04	
130.	0.80162D-05	0.74671D-04	
135.	0.80162D-05	0.74671D-04	
150.	0.80162D-05	0.74671D-04	
165.	0.80162D-05	0.74671D-04	
180.	0.80162D-05	0.74671D-04	

ALPHA = 0.02

THETA	RHO S	RHO P	RHO U
0.	0.20960D-03	0.20960D-03	0.10000D 01
5.	0.20959D-03	0.21120D-03	0.99241D 00
10.	0.20959D-03	0.21610D-03	0.96987D 00
20.	0.20957D-03	0.23732D-03	0.88310D 00
30.	0.20954D-03	0.27934D-03	0.75015D 00
40.	0.20949D-03	0.35690D-03	0.58706D 00
45.	0.20946D-03	0.41878D-03	0.50028D 00
50.	0.20943D-03	0.50664D-03	0.41349D 00
60.	0.20937D-03	0.83671D-03	0.25038D 00
70.	0.20929D-03	0.17856D-02	0.11740D 00
80.	0.20921D-03	0.68863D-02	0.30583D-01
90.	0.20913D-03	0.99993D 00	0.41818D-03
100.	0.20904D-03	0.68903D-02	0.30541D-01
110.	0.20896D-03	0.17839D-02	0.11732D 00
120.	0.20888D-03	0.83509D-03	0.25029D 00
130.	0.20881D-03	0.50525D-03	0.41340D 00
135.	0.20878D-03	0.41748D-03	0.50019D 00
150.	0.20870D-03	0.27825D-03	0.75010D 00
165.	0.20865D-03	0.22362D-03	0.93304D 00
180.	0.20863D-03	0.20863D-03	0.10000D 01

ALPHA = 0.04

THETA	RHO S	RHO P	RHO U
0.	0.21119D-03	0.21119D-03	0.10000D 01
5.	0.21119D-03	0.21280D-03	0.99241D 00
10.	0.21117D-03	0.21772D-03	0.96989D 00
20.	0.21108D-03	0.23901D-03	0.88317D 00
30.	0.21095D-03	0.28117D-03	0.75029D 00
40.	0.21076D-03	0.35895D-03	0.58726D 00
45.	0.21066D-03	0.42098D-03	0.50050D 00
50.	0.21054D-03	0.50904D-03	0.41372D 00
60.	0.21027D-03	0.83965D-03	0.25058D 00
70.	0.20997D-03	0.17892D-02	0.11754D 00
80.	0.20965D-03	0.68843D-02	0.30657D-01
90.	0.20932D-03	0.99973D 00	0.41860D-03
100.	0.20898D-03	0.69002D-02	0.30488D-01
110.	0.20865D-03	0.17824D-02	0.11724D 00
120.	0.20833D-03	0.83315D-03	0.25021D 00
130.	0.20804D-03	0.50347D-03	0.41334D 00
135.	0.20791D-03	0.41580D-03	0.50014D 00
150.	0.20759D-03	0.27678D-03	0.75008D 00
165.	0.20739D-03	0.22227D-03	0.93304D 00
180.	0.20732D-03	0.20732D-03	0.10000D 01

ALPHA = 0.06

THETA	RHO S	RHO P	RHO U
0.	0.21386D-03	0.21386D-03	0.10000D 01
5.	0.21384D-03	0.21547D-03	0.99242D 00
10.	0.21379D-03	0.22042D-03	0.96992D 00
20.	0.21361D-03	0.24184D-03	0.88330D 00
30.	0.21331D-03	0.28423D-03	0.75053D 00
40.	0.21289D-03	0.36237D-03	0.58759D 00
45.	0.21265D-03	0.42466D-03	0.50085D 00
50.	0.21238D-03	0.51304D-03	0.41408D 00
60.	0.21178D-03	0.84454D-03	0.25092D 00
70.	0.21111D-03	0.17951D-02	0.11779D 00
80.	0.21038D-03	0.68810D-02	0.30779D-01
90.	0.20963D-03	0.99938D 00	0.41930D-03
100.	0.20887D-03	0.69168D-02	0.30400D-01
110.	0.20812D-03	0.17799D-02	0.11711D 00
120.	0.20741D-03	0.82990D-03	0.25008D 00
130.	0.20676D-03	0.50050D-03	0.41324D 00
135.	0.20647D-03	0.41298D-03	0.50006D 00
150.	0.20574D-03	0.27432D-03	0.75005D 00
165.	0.20528D-03	0.22001D-03	0.93303D 00
180.	0.20512D-03	0.20512D-03	0.10000D 01

ALPHA = 0.08

THETA	RHO S	RHO P	RHO U
0.	0.21758D-03	0.21758D-03	0.10000D 01
5.	0.21755D-03	0.21921D-03	0.99244D 00
10.	0.21747D-03	0.22421D-03	0.96997D 00
20.	0.21714D-03	0.24579D-03	0.88347D 00
30.	0.21661D-03	0.28850D-03	0.75087D 00
40.	0.21587D-03	0.36715D-03	0.58806D 00
45.	0.21544D-03	0.42981D-03	0.50136D 00
50.	0.21496D-03	0.51864D-03	0.41460D 00
60.	0.21390D-03	0.85138D-03	0.25139D 00
70.	0.21270D-03	0.18034D-02	0.11814D 00
80.	0.21142D-03	0.68764D-02	0.30950D-01
90.	0.21007D-03	0.99890D 00	0.42029D-03
100.	0.20871D-03	0.69404D-02	0.30275D-01
110.	0.20738D-03	0.17764D-02	0.11692D 00
120.	0.20611D-03	0.82532D-03	0.24989D 00
130.	0.20495D-03	0.49629D-03	0.41309D 00
135.	0.20443D-03	0.40899D-03	0.49994D 00
150.	0.20312D-03	0.27085D-03	0.75000D 00
165.	0.20229D-03	0.21682D-03	0.93302D 00
180.	0.20201D-03	0.20201D-03	0.10000D 01

ALPHA = 0.02

THETA	RHO S	RHO P	RHO U
0.	0.77127D-03	0.77127D-03	0.10000D 01
5.	0.77127D-03	0.77717D-03	0.99242D 00
10.	0.77127D-03	0.79523D-03	0.96990D 00
20.	0.77127D-03	0.87333D-03	0.88323D 00
30.	0.77126D-03	0.10280D-02	0.75043D 00
40.	0.77125D-03	0.13134D-02	0.58753D 00
45.	0.77124D-03	0.15411D-02	0.50084D 00
50.	0.77124D-03	0.18642D-02	0.41415D 00
60.	0.77122D-03	0.30769D-02	0.25122D 00
70.	0.77119D-03	0.65518D-02	0.11839D 00
80.	0.77116D-03	0.24936D-01	0.31672D-01
90.	0.77113D-03	0.99993D 00	0.15411D-02
100.	0.77110D-03	0.24967D-01	0.31631D-01
110.	0.77106D-03	0.65548D-02	0.11831D 00
120.	0.77102D-03	0.30773D-02	0.25113D 00
130.	0.77099D-03	0.18640D-02	0.41406D 00
135.	0.77097D-03	0.15408D-02	0.50075D 00
150.	0.77093D-03	0.10276D-02	0.75038D 00
165.	0.77090D-03	0.82620D-03	0.93311D 00
180.	0.77089D-03	0.77089D-03	0.10000D 01

ALPHA = 0.04

THETA	RHO S	RHO P	RHO U
0.	0.77258D-03	0.77258D-03	0.10000D 01
5.	0.77258D-03	0.77849D-03	0.99242D 00
10.	0.77258D-03	0.79656D-03	0.96992D 00
20.	0.77256D-03	0.87472D-03	0.88330D 00
30.	0.77253D-03	0.10295D-02	0.75058D 00
40.	0.77249D-03	0.13151D-02	0.58773D 00
45.	0.77246D-03	0.15429D-02	0.50106D 00
50.	0.77243D-03	0.18661D-02	0.41438D 00
60.	0.77236D-03	0.30790D-02	0.25143D 00
70.	0.77226D-03	0.65527D-02	0.11854D 00
80.	0.77215D-03	0.24909D-01	0.31747D-01
90.	0.77202D-03	0.99973D 00	0.15431D-02
100.	0.77188D-03	0.25035D-01	0.31580D-01
110.	0.77174D-03	0.65648D-02	0.11824D 00
120.	0.77158D-03	0.30804D-02	0.25106D 00
130.	0.77144D-03	0.18654D-02	0.41400D 00
135.	0.77137D-03	0.15417D-02	0.50071D 00
150.	0.77120D-03	0.10280D-02	0.75036D 00
165.	0.77108D-03	0.82640D-03	0.93311D 00
180.	0.77104D-03	0.77104D-03	0.10000D 01

ALPHA = 0.06

THETA	RHO S	RHO P	RHO U
0.	0.77477D-03	0.77477D-03	0.10000D 01
5.	0.77477D-03	0.78068D-03	0.99243D 00
10.	0.77476D-03	0.79877D-03	0.96996D 00
20.	0.77472D-03	0.87704D-03	0.88343D 00
30.	0.77466D-03	0.10320D-02	0.75082D 00
40.	0.77456D-03	0.13179D-02	0.58806D 00
45.	0.77450D-03	0.15458D-02	0.50142D 00
50.	0.77443D-03	0.18693D-02	0.41475D 00
60.	0.77426D-03	0.30824D-02	0.25177D 00
70.	0.77405D-03	0.65541D-02	0.11878D 00
80.	0.77380D-03	0.24864D-01	0.31870D-01
90.	0.77351D-03	0.99938D 00	0.15463D-02
100.	0.77320D-03	0.25148D-01	0.31495D-01
110.	0.77286D-03	0.65815D-02	0.11811D 00
120.	0.77252D-03	0.30858D-02	0.25093D 00
130.	0.77220D-03	0.18677D-02	0.41391D 00
135.	0.77204D-03	0.15433D-02	0.50063D 00
150.	0.77165D-03	0.10287D-02	0.75033D 00
165.	0.77138D-03	0.82673D-03	0.93311D 00
180.	0.77129D-03	0.77129D-03	0.10000D 01

ALPHA = 0.08

THETA	RHO S	RHO P	RHO U
0.	0.77782D-03	0.77782D-03	0.10000D 01
5.	0.77782D-03	0.78374D-03	0.99244D 00
10.	0.77780D-03	0.80187D-03	0.97000D 00
20.	0.77774D-03	0.88028D-03	0.88360D 00
30.	0.77762D-03	0.10355D-02	0.75115D 00
40.	0.77746D-03	0.13217D-02	0.58852D 00
45.	0.77735D-03	0.15499D-02	0.50192D 00
50.	0.77723D-03	0.18737D-02	0.41527D 00
60.	0.77693D-03	0.30872D-02	0.25224D 00
70.	0.77656D-03	0.65562D-02	0.11913D 00
80.	0.77611D-03	0.24802D-01	0.32044D-01
90.	0.77561D-03	0.99891D 00	0.15509D-02
100.	0.77504D-03	0.25307D-01	0.31376D-01
110.	0.77445D-03	0.66051D-02	0.11793D 00
120.	0.77385D-03	0.30932D-02	0.25075D 00
130.	0.77327D-03	0.18709D-02	0.41377D 00
135.	0.77299D-03	0.15456D-02	0.50052D 00
150.	0.77228D-03	0.10296D-02	0.75029D 00
165.	0.77181D-03	0.82720D-03	0.93310D 00
180.	0.77165D-03	0.77165D-03	0.10000D 01

ALPHA = 0.02

THETA	RHO S	RHO P	RHO U
0.	0.15930D-02	0.15930D-02	0.10000D 01
5.	0.15930D-02	0.16052D-02	0.99243D 00
10.	0.15930D-02	0.16424D-02	0.96995D 00
20.	0.15930D-02	0.18036D-02	0.88342D 00
30.	0.15930D-02	0.21228D-02	0.75085D 00
40.	0.15931D-02	0.27114D-02	0.58821D 00
45.	0.15931D-02	0.31806D-02	0.50167D 00
50.	0.15931D-02	0.38463D-02	0.41512D 00
60.	0.15931D-02	0.63403D-02	0.25246D 00
70.	0.15932D-02	0.13451D-01	0.11984D 00
80.	0.15932D-02	0.50217D-01	0.33266D-01
90.	0.15932D-02	0.99993D 00	0.31815D-02
100.	0.15932D-02	0.50294D-01	0.33219D-01
110.	0.15932D-02	0.13462D-01	0.11976D 00
120.	0.15932D-02	0.63434D-02	0.25235D 00
130.	0.15932D-02	0.38476D-02	0.41502D 00
135.	0.15932D-02	0.31815D-02	0.50157D 00
150.	0.15932D-02	0.21232D-02	0.75079D 00
165.	0.15932D-02	0.17074D-02	0.93322D 00
180.	0.15932D-02	0.15932D-02	0.10000D 01

ALPHA = 0.04

THETA	RHO S	RHO P	RHO U
0.	0.15946D-02	0.15946D-02	0.10000D 01
5.	0.15946D-02	0.16068D-02	0.99244D 00
10.	0.15946D-02	0.16441D-02	0.96997D 00
20.	0.15947D-02	0.18053D-02	0.88350D 00
30.	0.15948D-02	0.21247D-02	0.75100D 00
40.	0.15949D-02	0.27134D-02	0.58843D 00
45.	0.15949D-02	0.31828D-02	0.50191D 00
50.	0.15950D-02	0.38486D-02	0.41537D 00
60.	0.15951D-02	0.63426D-02	0.25269D 00
70.	0.15952D-02	0.13450D-01	0.12001D 00
80.	0.15953D-02	0.50152D-01	0.33353D-01
90.	0.15954D-02	0.99973D 00	0.31862D-02
100.	0.15955D-02	0.50459D-01	0.33162D-01
110.	0.15955D-02	0.13491D-01	0.11967D 00
120.	0.15955D-02	0.63548D-02	0.25226D 00
130.	0.15955D-02	0.38537D-02	0.41494D 00
135.	0.15955D-02	0.31864D-02	0.50151D 00
150.	0.15954D-02	0.21262D-02	0.75076D 00
165.	0.15954D-02	0.17097D-02	0.93322D 00
180.	0.15953D-02	0.15953D-02	0.10000D 01

ALPHA = 0.06

THETA	RHO S	RHO P	RHO U
0.	0.15973D-02	0.15973D-02	0.10000D 01
5.	0.15973D-02	0.16095D-02	0.99245D 00
10.	0.15974D-02	0.16468D-02	0.97001D 00
20.	0.15975D-02	0.18082D-02	0.88364D 00
30.	0.15977D-02	0.21278D-02	0.75127D 00
40.	0.15979D-02	0.27169D-02	0.58880D 00
45.	0.15980D-02	0.31865D-02	0.50230D 00
50.	0.15982D-02	0.38525D-02	0.41578D 00
60.	0.15985D-02	0.63463D-02	0.25307D 00
70.	0.15987D-02	0.13448D-01	0.12029D 00
80.	0.15989D-02	0.50043D-01	0.33496D-01
90.	0.15991D-02	0.99938D 00	0.31941D-02
100.	0.15992D-02	0.50737D-01	0.33066D-01
110.	0.15993D-02	0.13541D-01	0.11952D 00
120.	0.15993D-02	0.63738D-02	0.25211D 00
130.	0.15992D-02	0.38640D-02	0.41482D 00
135.	0.15992D-02	0.31945D-02	0.50140D 00
150.	0.15991D-02	0.21312D-02	0.75071D 00
165.	0.15990D-02	0.17136D-02	0.93321D 00
180.	0.15989D-02	0.15989D-02	0.10000D 01

ALPHA = 0.08

THETA	RHO S	RHO P	RHO U
0.	0.16011D-02	0.16011D-02	0.10000D 01
5.	0.16011D-02	0.16133D-02	0.99246D 00
10.	0.16011D-02	0.16506D-02	0.97006D 00
20.	0.16014D-02	0.18122D-02	0.88383D 00
30.	0.16017D-02	0.21321D-02	0.75163D 00
40.	0.16021D-02	0.27217D-02	0.58931D 00
45.	0.16024D-02	0.31916D-02	0.50286D 00
50.	0.16026D-02	0.38578D-02	0.41635D 00
60.	0.16031D-02	0.63515D-02	0.25360D 00
70.	0.16036D-02	0.13445D-01	0.12068D 00
80.	0.16040D-02	0.49893D-01	0.33698D-01
90.	0.16043D-02	0.99891D 00	0.32052D-02
100.	0.16045D-02	0.51130D-01	0.32933D-01
110.	0.16046D-02	0.13610D-01	0.11931D 00
120.	0.16046D-02	0.64007D-02	0.25189D 00
130.	0.16045D-02	0.38785D-02	0.41464D 00
135.	0.16045D-02	0.32060D-02	0.50125D 00
150.	0.16043D-02	0.21383D-02	0.75065D 00
165.	0.16041D-02	0.17191D-02	0.93319D 00
180.	0.16040D-02	0.16040D-02	0.10000D 01

ALPHA = 0.02

THETA	RHO S	RHO P	RHO U
0.	0.25939D-02	0.25939D-02	0.10000D 01
5.	0.25939D-02	0.26137D-02	0.99245D 00
10.	0.25939D-02	0.26744D-02	0.97001D 00
20.	0.25940D-02	0.29365D-02	0.88366D 00
30.	0.25940D-02	0.34555D-02	0.75135D 00
40.	0.25941D-02	0.44119D-02	0.58904D 00
45.	0.25941D-02	0.51739D-02	0.50267D 00
50.	0.25942D-02	0.62542D-02	0.41630D 00
60.	0.25943D-02	0.10293D-01	0.25396D 00
70.	0.25943D-02	0.21741D-01	0.12161D 00
80.	0.25944D-02	0.79338D-01	0.35204D-01
90.	0.25945D-02	0.99993D 00	0.51757D-02
100.	0.25945D-02	0.79474D-01	0.35150D-01
110.	0.25946D-02	0.21761D-01	0.12151D 00
120.	0.25946D-02	0.10300D-01	0.25384D 00
130.	0.25946D-02	0.62572D-02	0.41618D 00
135.	0.25946D-02	0.51761D-02	0.50256D 00
150.	0.25947D-02	0.34566D-02	0.75128D 00
165.	0.25947D-02	0.27804D-02	0.93336D 00
180.	0.25947D-02	0.25947D-02	0.10000D 01

ALPHA = 0.04

THETA	RHO S	RHO P	RHO U
0.	0.25965D-02	0.25965D-02	0.10000D 01
5.	0.25965D-02	0.26163D-02	0.99245D 00
10.	0.25965D-02	0.26769D-02	0.97004D 00
20.	0.25966D-02	0.29392D-02	0.88375D 00
30.	0.25968D-02	0.34584D-02	0.75152D 00
40.	0.25971D-02	0.44152D-02	0.58928D 00
45.	0.25972D-02	0.51775D-02	0.50294D 00
50.	0.25974D-02	0.62579D-02	0.41657D 00
60.	0.25977D-02	0.10297D-01	0.25422D 00
70.	0.25980D-02	0.21738D-01	0.12180D 00
80.	0.25983D-02	0.79224D-01	0.35304D-01
90.	0.25986D-02	0.99973D 00	0.51845D-02
100.	0.25988D-02	0.79771D-01	0.35087D-01
110.	0.25990D-02	0.21818D-01	0.12141D 00
120.	0.25992D-02	0.10323D-01	0.25373D 00
130.	0.25993D-02	0.62698D-02	0.41609D 00
135.	0.25993D-02	0.51863D-02	0.50248D 00
150.	0.25993D-02	0.34630D-02	0.75124D 00
165.	0.25993D-02	0.27855D-02	0.93335D 00
180.	0.25993D-02	0.25993D-02	0.10000D 01

ALPHA = 0.06

THETA	RHO S	RHO P	RHO U
0.	0.26007D-02	0.26007D-02	0.10000D 01
5.	0.26007D-02	0.26205D-02	0.99246D 00
10.	0.26008D-02	0.26812D-02	0.97008D 00
20.	0.26011D-02	0.29437D-02	0.88390D 00
30.	0.26015D-02	0.34633D-02	0.75181D 00
40.	0.26021D-02	0.44207D-02	0.58968D 00
45.	0.26024D-02	0.51833D-02	0.50337D 00
50.	0.26028D-02	0.62641D-02	0.41703D 00
60.	0.26035D-02	0.10303D-01	0.25464D 00
70.	0.26042D-02	0.21733D-01	0.12211D 00
80.	0.26049D-02	0.79036D-01	0.35471D-01
90.	0.26055D-02	0.99939D 00	0.51991D-02
100.	0.26061D-02	0.80269D-01	0.34982D-01
110.	0.26065D-02	0.21913D-01	0.12124D 00
120.	0.26068D-02	0.10361D-01	0.25355D 00
130.	0.26070D-02	0.62908D-02	0.41593D 00
135.	0.26070D-02	0.52032D-02	0.50235D 00
150.	0.26072D-02	0.34738D-02	0.75118D 00
165.	0.26072D-02	0.27940D-02	0.93333D 00
180.	0.26072D-02	0.26072D-02	0.10000D 01

ALPHA = 0.08

THETA	RHO S	RHO P	RHO U
0.	0.26065D-02	0.26065D-02	0.10000D 01
5.	0.26066D-02	0.26264D-02	0.99248D 00
10.	0.26067D-02	0.26872D-02	0.97013D 00
20.	0.26072D-02	0.29500D-02	0.88411D 00
30.	0.26080D-02	0.34702D-02	0.75221D 00
40.	0.26091D-02	0.44284D-02	0.59024D 00
45.	0.26097D-02	0.51914D-02	0.50398D 00
50.	0.26103D-02	0.62726D-02	0.41766D 00
60.	0.26116D-02	0.10311D-01	0.25523D 00
70.	0.26129D-02	0.21726D-01	0.12256D 00
80.	0.26141D-02	0.78775D-01	0.35705D-01
90.	0.26152D-02	0.99891D 00	0.52196D-02
100.	0.26162D-02	0.80973D-01	0.34834D-01
110.	0.26169D-02	0.22047D-01	0.12100D 00
120.	0.26175D-02	0.10414D-01	0.25329D 00
130.	0.26178D-02	0.63205D-02	0.41571D 00
135.	0.26180D-02	0.52270D-02	0.50216D 00
150.	0.26182D-02	0.34889D-02	0.75109D 00
165.	0.26183D-02	0.28059D-02	0.93331D 00
180.	0.26183D-02	0.26183D-02	0.10000D 01

ALPHA = 0.02

THETA	RHO S	RHO P	RHO U
0.	0.24429D-03	0.24429D-03	0.10000D 01
5.	0.24430D-03	0.24617D-03	0.99242D 00
10.	0.24432D-03	0.25191D-03	0.96991D 00
20.	0.24441D-03	0.27673D-03	0.88324D 00
30.	0.24456D-03	0.32592D-03	0.75043D 00
40.	0.24476D-03	0.41672D-03	0.58744D 00
45.	0.24487D-03	0.48919D-03	0.50069D 00
50.	0.24500D-03	0.59211D-03	0.41392D 00
60.	0.24527D-03	0.97878D-03	0.25077D 00
70.	0.24557D-03	0.20903D-02	0.11770D 00
80.	0.24587D-03	0.80546D-02	0.30764D-01
90.	0.24618D-03	0.99945D 00	0.49238D-03
100.	0.24649D-03	0.81320D-02	0.30549D-01
110.	0.24677D-03	0.21075D-02	0.11731D 00
120.	0.24703D-03	0.98769D-03	0.25029D 00
130.	0.24726D-03	0.59827D-03	0.41344D 00
135.	0.24736D-03	0.49461D-03	0.50024D 00
150.	0.24761D-03	0.33011D-03	0.75015D 00
165.	0.24776D-03	0.26554D-03	0.93306D 00
180.	0.24781D-03	0.24781D-03	0.10000D 01

ALPHA = 0.04

THETA	RHO S	RHO P	RHO U
0.	0.23900D-03	0.23900D-03	0.10000D 01
5.	0.23903D-03	0.24085D-03	0.99245D 00
10.	0.23912D-03	0.24651D-03	0.97004D 00
20.	0.23949D-03	0.27101D-03	0.88372D 00
30.	0.24007D-03	0.31955D-03	0.75134D 00
40.	0.24086D-03	0.40922D-03	0.58869D 00
45.	0.24132D-03	0.48081D-03	0.50203D 00
50.	0.24182D-03	0.58250D-03	0.41528D 00
60.	0.24291D-03	0.96467D-03	0.25199D 00
70.	0.24410D-03	0.20627D-02	0.11855D 00
80.	0.24534D-03	0.79319D-02	0.31168D-01
90.	0.24658D-03	0.99779D 00	0.49359D-03
100.	0.24780D-03	0.82430D-02	0.30303D-01
110.	0.24896D-03	0.21318D-02	0.11700D 00
120.	0.25001D-03	0.10005D-02	0.25007D 00
130.	0.25095D-03	0.60731D-03	0.41335D 00
135.	0.25136D-03	0.50262D-03	0.50022D 00
150.	0.25237D-03	0.33641D-03	0.75023D 00
165.	0.25298D-03	0.27113D-03	0.93310D 00
180.	0.25319D-03	0.25319D-03	0.10000D 01

P =   0.20   M = 1.05

ALPHA = 0.06

THETA	RHO S	RHO P	RHO U
0.	0.23015D-03	0.23015D-03	0.10000D 01
5.	0.23022D-03	0.23196D-03	0.99251D 00
10.	0.23042D-03	0.23749D-03	0.97026D 00
20.	0.23123D-03	0.26143D-03	0.88453D 00
30.	0.23255D-03	0.30891D-03	0.75286D 00
40.	0.23432D-03	0.39668D-03	0.59079D 00
45.	0.23535D-03	0.46681D-03	0.50429D 00
50.	0.23647D-03	0.56647D-03	0.41759D 00
60.	0.23893D-03	0.94116D-03	0.254405D 00
70.	0.24162D-03	0.20170D-02	0.12000D 00
80.	0.24442D-03	0.77308D-02	0.31854D-01
90.	0.24726D-03	0.99500D 00	0.49565D-03
100.	0.25005D-03	0.84365D-02	0.29881D-01
110.	0.25269D-03	0.21737D-02	0.11647D 00
120.	0.25513D-03	0.10226D-02	0.24968D 00
130.	0.25728D-03	0.62287D-03	0.41321D 00
135.	0.25824D-03	0.51642D-03	0.50019D 00
150.	0.26058D-03	0.34729D-03	0.75037D 00
165.	0.26202D-03	0.28079D-03	0.93317D 00
180.	0.26251D-03	0.26251D-03	0.10000D 01

ALPHA = 0.08

THETA	RHO S	RHO P	RHO U
0.	0.21768D-03	0.21768D-03	0.10000D 01
5.	0.21780D-03	0.21943D-03	0.99259D 00
10.	0.21816D-03	0.22478D-03	0.97058D 00
20.	0.21959D-03	0.24794D-03	0.88567D 00
30.	0.22191D-03	0.29394D-03	0.75502D 00
40.	0.22504D-03	0.37907D-03	0.59376D 00
45.	0.22688D-03	0.44716D-03	0.50748D 00
50.	0.22887D-03	0.54399D-03	0.42086D 00
60.	0.23326D-03	0.90830D-03	0.25698D 00
70.	0.23806D-03	0.19534D-02	0.12208D 00
80.	0.24311D-03	0.74565D-02	0.32839D-01
90.	0.24825D-03	0.99101D 00	0.49862D-03
100.	0.25331D-03	0.87275D-02	0.29270D-01
110.	0.25814D-03	0.22355D-02	0.11570D 00
120.	0.26261D-03	0.10550D-02	0.24912D 00
130.	0.26660D-03	0.64575D-03	0.41300D 00
135.	0.26837D-03	0.53672D-03	0.50015D 00
150.	0.27271D-03	0.36337D-03	0.75058D 00
165.	0.27541D-03	0.29510D-03	0.93328D 00
180.	0.27632D-03	0.27632D-03	0.10000D 01

ALPHA = 0.02

THETA	RHO S	RHO P	RHO U
0.	0.92135D-03	0.92135D-03	0.10000D 01
5.	0.92138D-03	0.92841D-03	0.99243D 00
10.	0.92145D-03	0.95002D-03	0.96995D 00
20.	0.92174D-03	0.10435D-02	0.88341D 00
30.	0.92221D-03	0.12287D-02	0.75079D 00
40.	0.92284D-03	0.15704D-02	0.58804D 00
45.	0.92321D-03	0.18429D-02	0.50141D 00
50.	0.92361D-03	0.22297D-02	0.41476D 00
60.	0.92447D-03	0.36810D-02	0.25184D 00
70.	0.92541D-03	0.78347D-02	0.11893D 00
80.	0.92638D-03	0.29686D-01	0.32103D-01
90.	0.92735D-03	0.99945D 00	0.18535D-02
100.	0.92829D-03	0.29989D-01	0.31853D-01
110.	0.92917D-03	0.78968D-02	0.11848D 00
120.	0.92997D-03	0.37112D-02	0.25128D 00
130.	0.93067D-03	0.22499D-02	0.41420D 00
135.	0.93098D-03	0.18604D-02	0.50089D 00
150.	0.93173D-03	0.12419D-02	0.75047D 00
165.	0.93219D-03	0.99904D-03	0.93315D 00
180.	0.93234D-03	0.93234D-03	0.10000D 01

ALPHA = 0.04

THETA	RHO S	RHO P	RHO U
0.	0.90512D-03	0.90512D-03	0.10000D 01
5.	0.90522D-03	0.91210D-03	0.99247D 00
10.	0.90551D-03	0.93345D-03	0.97010D 00
20.	0.90668D-03	0.10258D-02	0.88394D 00
30.	0.90856D-03	0.12089D-02	0.75178D 00
40.	0.91107D-03	0.15467D-02	0.58941D 00
45.	0.91254D-03	0.18163D-02	0.50288D 00
50.	0.91413D-03	0.21988D-02	0.41626D 00
60.	0.91760D-03	0.36340D-02	0.25319D 00
70.	0.92135D-03	0.77370D-02	0.11989D 00
80.	0.92524D-03	0.29215D-01	0.32565D-01
90.	0.92914D-03	0.99780D 00	0.18586D-02
100.	0.93294D-03	0.30434D-01	0.31558D-01
110.	0.93650D-03	0.79863D-02	0.11809D 00
120.	0.93975D-03	0.37553D-02	0.25095D 00
130.	0.94261D-03	0.22798D-02	0.41402D 00
135.	0.94387D-03	0.18866D-02	0.50078D 00
150.	0.94691D-03	0.12621D-02	0.75050D 00
165.	0.94878D-03	0.10168D-02	0.93317D 00
180.	0.94941D-03	0.94941D-03	0.10000D 01

ALPHA = 0.06

THETA	RHO S	RHO P	RHO U
0.	0.87798D-03	0.87798D-03	0.10000D 01
5.	0.87820D-03	0.88481D-03	0.99253D 00
10.	0.87885D-03	0.90574D-03	0.97034D 00
20.	0.88145D-03	0.99631D-03	0.88482D 00
30.	0.88566D-03	0.11758D-02	0.75345D 00
40.	0.89132D-03	0.15073D-02	0.59170D 00
45.	0.89462D-03	0.17718D-02	0.50535D 00
50.	0.89819D-03	0.21474D-02	0.41879D 00
60.	0.90601D-03	0.35558D-02	0.25547D 00
70.	0.91448D-03	0.75754D-02	0.12152D 00
80.	0.92331D-03	0.28447D-01	0.33350D-01
90.	0.93220D-03	0.99501D 00	0.18673D-02
100.	0.94086D-03	0.31212D-01	0.31056D-01
110.	0.94903D-03	0.81408D-02	0.11742D 00
120.	0.95651D-03	0.38311D-02	0.25039D 00
130.	0.96309D-03	0.23312D-02	0.41370D 00
135.	0.96601D-03	0.19316D-02	0.50059D 00
150.	0.97307D-03	0.12969D-02	0.75054D 00
165.	0.97741D-03	0.10474D-02	0.93321D 00
180.	0.97887D-03	0.97887D-03	0.10000D 01

ALPHA = 0.08

THETA	RHO S	RHO P	RHO U
0.	0.83977D-03	0.83977D-03	0.10000D 01
5.	0.84016D-03	0.84641D-03	0.99262D 00
10.	0.84132D-03	0.86675D-03	0.97068D 00
20.	0.84590D-03	0.95479D-03	0.88606D 00
30.	0.85334D-03	0.11294D-02	0.75579D 00
40.	0.86336D-03	0.14520D-02	0.59494D 00
45.	0.86921D-03	0.17096D-02	0.50885D 00
50.	0.87556D-03	0.20754D-02	0.42238D 00
60.	0.88950D-03	0.34469D-02	0.25872D 00
70.	0.90467D-03	0.73519D-02	0.12385D 00
80.	0.92055D-03	0.27408D-01	0.34475D-01
90.	0.93661D-03	0.99105D 00	0.18799D-02
100.	0.95233D-03	0.32385D-01	0.30330D-01
110.	0.96725D-03	0.83686D-02	0.11644D 00
120.	0.98096D-03	0.39424D-02	0.24956D 00
130.	0.99310D-03	0.24066D-02	0.41323D 00
135.	0.99849D-03	0.19978D-02	0.50030D 00
150.	0.10116D-02	0.13482D-02	0.75061D 00
165.	0.10197D-02	0.10927D-02	0.93328D 00
180.	0.10224D-02	0.10224D-02	0.10000D 01

ALPHA = 0.02

THETA	RHO S	RHO P	RHO U
0.	0.32695D-02	0.32695D-02	0.10000D 01
5.	0.32696D-02	0.32945D-02	0.99247D 00
10.	0.32697D-02	0.33709D-02	0.97010D 00
20.	0.32704D-02	0.37013D-02	0.88398D 00
30.	0.32715D-02	0.43552D-02	0.75199D 00
40.	0.32730D-02	0.55600D-02	0.59002D 00
45.	0.32739D-02	0.65194D-02	0.50380D 00
50.	0.32748D-02	0.78787D-02	0.41756D 00
60.	0.32768D-02	0.12954D-01	0.25540D 00
70.	0.32789D-02	0.27271D-01	0.12311D 00
80.	0.32810D-02	0.97912D-01	0.36671D-01
90.	0.32831D-02	0.99945D 00	0.65466D-02
100.	0.32851D-02	0.98878D-01	0.36390D-01
110.	0.32870D-02	0.27456D-01	0.12260D 00
120.	0.32886D-02	0.13033D-01	0.25477D 00
130.	0.32900D-02	0.79275D-02	0.41693D 00
135.	0.32906D-02	0.65605D-02	0.50321D 00
150.	0.32921D-02	0.43847D-02	0.75163D 00
165.	0.32929D-02	0.35285D-02	0.93346D 00
180.	0.32932D-02	0.32932D-02	0.10000D 01

ALPHA = 0.04

THETA	RHO S	RHO P	RHO U
0.	0.32368D-02	0.32368D-02	0.10000D 01
5.	0.32370D-02	0.32615D-02	0.99251D 00
10.	0.32377D-02	0.33373D-02	0.97025D 00
20.	0.32405D-02	0.36650D-02	0.88454D 00
30.	0.32449D-02	0.43135D-02	0.75307D 00
40.	0.32508D-02	0.55082D-02	0.59150D 00
45.	0.32542D-02	0.64594D-02	0.50540D 00
50.	0.32578D-02	0.78068D-02	0.41920D 00
60.	0.32658D-02	0.12834D-01	0.25688D 00
70.	0.32742D-02	0.26991D-01	0.12418D 00
80.	0.32829D-02	0.96449D-01	0.37198D-01
90.	0.32914D-02	0.99780D 00	0.65684D-02
100.	0.32995D-02	0.10032D 00	0.36069D-01
110.	0.33069D-02	0.27731D-01	0.12215D 00
120.	0.33135D-02	0.13154D-01	0.25437D 00
130.	0.33192D-02	0.80031D-02	0.41668D 00
135.	0.33217D-02	0.66249D-02	0.50304D 00
150.	0.33275D-02	0.44320D-02	0.75162D 00
165.	0.33310D-02	0.35693D-02	0.93347D 00
180.	0.33322D-02	0.33322D-02	0.10000D 01

ALPHA = 0.06

THETA	RHO S	RHO P	RHO U
0.	0.31821D-02	0.31821D-02	0.10000D 01
5.	0.31826D-02	0.32065D-02	0.99257D 00
10.	0.31842D-02	0.32813D-02	0.97051D 00
20.	0.31904D-02	0.36044D-02	0.88549D 00
30.	0.32003D-02	0.42440D-02	0.75486D 00
40.	0.32135D-02	0.54220D-02	0.59399D 00
45.	0.32212D-02	0.63597D-02	0.50808D 00
50.	0.32294D-02	0.76875D-02	0.42195D 00
60.	0.32473D-02	0.12637D-01	0.25938D 00
70.	0.32664D-02	0.26529D-01	0.12598D 00
80.	0.32860D-02	0.94077D-01	0.38090D-01
90.	0.33053D-02	0.99503D 00	0.66054D-02
100.	0.33238D-02	0.10285D 00	0.35524D-01
110.	0.33408D-02	0.28204D-01	0.12139D 00
120.	0.33560D-02	0.13361D-01	0.25369D 00
130.	0.33691D-02	0.81325D-02	0.41625D 00
135.	0.33748D-02	0.67353D-02	0.50274D 00
150.	0.33884D-02	0.45132D-02	0.75161D 00
165.	0.33965D-02	0.36394D-02	0.93350D 00
180.	0.33992D-02	0.33992D-02	0.10000D 01

ALPHA = 0.08

THETA	RHO S	RHO P	RHO U
0.	0.31054D-02	0.31054D-02	0.10000D 01
5.	0.31063D-02	0.31293D-02	0.99267D 00
10.	0.31091D-02	0.32026D-02	0.97088D 00
20.	0.31199D-02	0.35195D-02	0.88682D 00
30.	0.31375D-02	0.41467D-02	0.75739D 00
40.	0.31609D-02	0.53016D-02	0.59749D 00
45.	0.31745D-02	0.62206D-02	0.51187D 00
50.	0.31892D-02	0.75214D-02	0.42585D 00
60.	0.32211D-02	0.12362D-01	0.26293D 00
70.	0.32553D-02	0.25895D-01	0.12855D 00
80.	0.32905D-02	0.90889D-01	0.39364D-01
90.	0.33254D-02	0.99113D 00	0.66584D-02
100.	0.33589D-02	0.10664D 00	0.34738D-01
110.	0.33899D-02	0.28901D-01	0.12028D 00
120.	0.34178D-02	0.13663D-01	0.25270D 00
130.	0.34419D-02	0.83217D-02	0.41562D 00
135.	0.34524D-02	0.68967D-02	0.50231D 00
150.	0.34775D-02	0.46323D-02	0.75158D 00
165.	0.34927D-02	0.37423D-02	0.93354D 00
180.	0.34977D-02	0.34977D-02	0.10000D 01

ALPHA = 0.02

THETA	RHO S	RHO P	RHO U
0.	0.65135D-02	0.65135D-02	0.10000D 01
5.	0.65136D-02	0.65630D-02	0.99252D 00
10.	0.65138D-02	0.67146D-02	0.97029D 00
20.	0.65149D-02	0.73699D-02	0.88474D 00
30.	0.65165D-02	0.86654D-02	0.75362D 00
40.	0.65186D-02	0.11048D-01	0.59270D 00
45.	0.65198D-02	0.12941D-01	0.50705D 00
50.	0.65212D-02	0.15616D-01	0.42136D 00
60.	0.65240D-02	0.25542D-01	0.26025D 00
70.	0.65269D-02	0.53011D-01	0.12881D 00
80.	0.65298D-02	0.17805D 00	0.42923D-01
90.	0.65327D-02	0.99945D 00	0.12984D-01
100.	0.65353D-02	0.17969D 00	0.42626D-01
110.	0.65376D-02	0.53335D-01	0.12828D 00
120.	0.65396D-02	0.25671D-01	0.25959D 00
130.	0.65412D-02	0.15690D-01	0.42070D 00
135.	0.65419D-02	0.13001D-01	0.50642D 00
150.	0.65435D-02	0.87059D-02	0.75324D 00
165.	0.65444D-02	0.70110D-02	0.93389D 00
180.	0.65447D-02	0.65447D-02	0.10000D 01

ALPHA = 0.04

THETA	RHO S	RHO P	RHO U
0.	0.64755D-02	0.64755D-02	0.10000D 01
5.	0.64758D-02	0.65247D-02	0.99256D 00
10.	0.64768D-02	0.66753D-02	0.97046D 00
20.	0.64809D-02	0.73264D-02	0.88533D 00
30.	0.64874D-02	0.86136D-02	0.75475D 00
40.	0.64959D-02	0.10980D-01	0.59426D 00
45.	0.65008D-02	0.12859D-01	0.50873D 00
50.	0.65061D-02	0.15515D-01	0.42310D 00
60.	0.65173D-02	0.25357D-01	0.26183D 00
70.	0.65291D-02	0.52537D-01	0.12996D 00
80.	0.65409D-02	0.17560D 00	0.43506D-01
90.	0.65523D-02	0.99780D 00	0.13034D-01
100.	0.65628D-02	0.18218D 00	0.42308D-01
110.	0.65722D-02	0.53835D-01	0.12781D 00
120.	0.65803D-02	0.25877D-01	0.25917D 00
130.	0.65870D-02	0.15811D-01	0.42042D 00
135.	0.65898D-02	0.13102D-01	0.50623D 00
150.	0.65963D-02	0.87764D-02	0.75322D 00
165.	0.66000D-02	0.70704D-02	0.93390D 00
180.	0.66012D-02	0.66012D-02	0.10000D 01

ALPHA = 0.06

THETA	RHO S	RHO P	RHO U
0.	0.64121D-02	0.64121D-02	0.10000D 01
5.	0.64129D-02	0.64608D-02	0.99263D 00
10.	0.64152D-02	0.66099D-02	0.97073D 00
20.	0.64242D-02	0.72541D-02	0.88633D 00
30.	0.64388D-02	0.85274D-02	0.75663D 00
40.	0.64580D-02	0.10867D-01	0.59688D 00
45.	0.64690D-02	0.12724D-01	0.51156D 00
50.	0.64808D-02	0.15347D-01	0.42600D 00
60.	0.65062D-02	0.25054D-01	0.26448D 00
70.	0.65329D-02	0.51759D-01	0.13189D 00
80.	0.65596D-02	0.17163D 00	0.44487D-01
90.	0.65854D-02	0.99506D 00	0.13117D-01
100.	0.66094D-02	0.18651D 00	0.41771D-01
110.	0.66309D-02	0.54694D-01	0.12703D 00
120.	0.66495D-02	0.26228D-01	0.25845D 00
130.	0.66649D-02	0.16018D-01	0.41996D 00
135.	0.66715D-02	0.13274D-01	0.50590D 00
150.	0.66865D-02	0.88972D-02	0.75318D 00
165.	0.66952D-02	0.71723D-02	0.93392D 00
180.	0.66980D-02	0.66980D-02	0.10000D 01

ALPHA = 0.08

THETA	RHO S	RHO P	RHO U
0.	0.63235D-02	0.63235D-02	0.10000D 01
5.	0.63249D-02	0.63715D-02	0.99273D 00
10.	0.63289D-02	0.65184D-02	0.97112D 00
20.	0.63449D-02	0.71532D-02	0.88772D 00
30.	0.63706D-02	0.84073D-02	0.75927D 00
40.	0.64046D-02	0.10710D-01	0.60055D 00
45.	0.64242D-02	0.12537D-01	0.51553D 00
50.	0.64452D-02	0.15115D-01	0.43009D 00
60.	0.64905D-02	0.24634D-01	0.26822D 00
70.	0.65381D-02	0.50695D-01	0.13463D 00
80.	0.65862D-02	0.16632D 00	0.45885D-01
90.	0.66328D-02	0.99120D 00	0.13237D-01
100.	0.66763D-02	0.19297D 00	0.41001D-01
110.	0.67155D-02	0.55952D-01	0.12589D 00
120.	0.67495D-02	0.26740D-01	0.25742D 00
130.	0.67780D-02	0.16319D-01	0.41928D 00
135.	0.67901D-02	0.13524D-01	0.50543D 00
150.	0.68180D-02	0.90732D-02	0.75313D 00
165.	0.68342D-02	0.73210D-02	0.93395D 00
180.	0.68394D-02	0.68394D-02	0.10000D 01

ALPHA = 0.02

THETA	RHO S	RHO P	RHO U
0.	0.10255D-01	0.10255D-01	0.10000D 01
5.	0.10256D-01	0.10333D-01	0.99257D 00
10.	0.10256D-01	0.10571D-01	0.97052D 00
20.	0.10257D-01	0.11597D-01	0.88561D 00
30.	0.10259D-01	0.13625D-01	0.75547D 00
40.	0.10262D-01	0.17345D-01	0.59577D 00
45.	0.10263D-01	0.20294D-01	0.51075D 00
50.	0.10265D-01	0.24451D-01	0.42571D 00
60.	0.10268D-01	0.39759D-01	0.26580D 00
70.	0.10272D-01	0.81230D-01	0.13533D 00
80.	0.10275D-01	0.25485D 00	0.50077D-01
90.	0.10278D-01	0.99945D 00	0.20352D-01
100.	0.10281D-01	0.25704D 00	0.49765D-01
110.	0.10283D-01	0.81694D-01	0.13477D 00
120.	0.10285D-01	0.39935D-01	0.26510D 00
130.	0.10287D-01	0.24545D-01	0.42501D 00
135.	0.10287D-01	0.20368D-01	0.51010D 00
150.	0.10288D-01	0.13671D-01	0.75507D 00
165.	0.10289D-01	0.11020D-01	0.93438D 00
180.	0.10289D-01	0.10289D-01	0.10000D 01

ALPHA = 0.04

THETA	RHO S	RHO P	RHO U
0.	0.10223D-01	0.10223D-01	0.10000D 01
5.	0.10223D-01	0.10300D-01	0.99262D 00
10.	0.10225D-01	0.10537D-01	0.97069D 00
20.	0.10230D-01	0.11558D-01	0.88623D 00
30.	0.10237D-01	0.13575D-01	0.75665D 00
40.	0.10248D-01	0.17274D-01	0.59740D 00
45.	0.10254D-01	0.20204D-01	0.51252D 00
50.	0.10260D-01	0.24333D-01	0.42753D 00
60.	0.10274D-01	0.39523D-01	0.26746D 00
70.	0.10287D-01	0.80575D-01	0.13656D 00
80.	0.10301D-01	0.25161D 00	0.50718D-01
90.	0.10313D-01	0.99781D 00	0.20438D-01
100.	0.10324D-01	0.26038D 00	0.49464D-01
110.	0.10334D-01	0.82432D-01	0.13431D 00
120.	0.10342D-01	0.40230D-01	0.26468D 00
130.	0.10348D-01	0.24710D-01	0.42473D 00
135.	0.10350D-01	0.20503D-01	0.50989D 00
150.	0.10356D-01	0.13761D-01	0.75505D 00
165.	0.10358D-01	0.11094D-01	0.93439D 00
180.	0.10359D-01	0.10359D-01	0.10000D 01

ALPHA = 0.06

THETA	RHO S	RHO P	RHO U
0.	0.10169D-01	0.10169D-01	0.10000D 01
5.	0.10170D-01	0.10245D-01	0.99269D 00
10.	0.10173D-01	0.10480D-01	0.97097D 00
20.	0.10184D-01	0.11493D-01	0.88726D 00
30.	0.10202D-01	0.13491D-01	0.75861D 00
40.	0.10225D-01	0.17155D-01	0.60012D 00
45.	0.10238D-01	0.20055D-01	0.51546D 00
50.	0.10252D-01	0.24139D-01	0.43056D 00
60.	0.10283D-01	0.39136D-01	0.27025D 00
70.	0.10314D-01	0.79504D-01	0.13861D 00
80.	0.10344D-01	0.24637D 00	0.51794D-01
90.	0.10373D-01	0.99508D 00	0.20583D-01
100.	0.10398D-01	0.26616D 00	0.48956D-01
110.	0.10420D-01	0.83697D-01	0.13353D 00
120.	0.10438D-01	0.40731D-01	0.26395D 00
130.	0.10452D-01	0.24992D-01	0.42425D 00
135.	0.10458D-01	0.20733D-01	0.50955D 00
150.	0.10470D-01	0.13915D-01	0.75501D 00
165.	0.10477D-01	0.11220D-01	0.93441D 00
180.	0.10479D-01	0.10479D-01	0.10000D 01

ALPHA = 0.08

THETA	RHO S	RHO P	RHO U
0.	0.10093D-01	0.10093D-01	0.10000D 01
5.	0.10095D-01	0.10169D-01	0.99279D 00
10.	0.10100D-01	0.10401D-01	0.97137D 00
20.	0.10120D-01	0.11402D-01	0.88870D 00
30.	0.10151D-01	0.13376D-01	0.76135D 00
40.	0.10193D-01	0.16991D-01	0.60393D 00
45.	0.10216D-01	0.19850D-01	0.51959D 00
50.	0.10242D-01	0.23871D-01	0.43482D 00
60.	0.10295D-01	0.38603D-01	0.27417D 00
70.	0.10351D-01	0.78047D-01	0.14151D 00
80.	0.10406D-01	0.23938D 00	0.53320D-01
90.	0.10457D-01	0.99127D 00	0.20789D-01
100.	0.10504D-01	0.27472D 00	0.48231D-01
110.	0.10544D-01	0.85540D-01	0.13241D 00
120.	0.10577D-01	0.41459D-01	0.26291D 00
130.	0.10603D-01	0.25400D-01	0.42355D 00
135.	0.10614D-01	0.21065D-01	0.50906D 00
150.	0.10637D-01	0.14138D-01	0.75494D 00
165.	0.10649D-01	0.11405D-01	0.93444D 00
180.	0.10653D-01	0.10653D-01	0.10000D 01

ALPHA = 0.02

THETA	RHO S	RHO P	RHO U
0.	0.36802D-03	0.36802D-03	0.10000D 01
5.	0.36810D-03	0.37090D-03	0.99246D 00
10.	0.36836D-03	0.37972D-03	0.97007D 00
20.	0.36935D-03	0.41791D-03	0.88385D 00
30.	0.37097D-03	0.49364D-03	0.75158D 00
40.	0.37315D-03	0.63363D-03	0.58906D 00
45.	0.37443D-03	0.74549D-03	0.50245D 00
50.	0.37583D-03	0.90445D-03	0.41575D 00
60.	0.37889D-03	0.15023D-02	0.25249D 00
70.	0.38225D-03	0.32205D-02	0.11903D 00
80.	0.38579D-03	0.12368D-01	0.31567D-01
90.	0.38940D-03	0.99772D 00	0.77939D-03
100.	0.39297D-03	0.13062D-01	0.30466D-01
110.	0.39638D-03	0.33963D-02	0.11706D 00
120.	0.39954D-03	0.15998D-02	0.25005D 00
130.	0.40237D-03	0.97413D-03	0.41329D 00
135.	0.40363D-03	0.80735D-03	0.50015D 00
150.	0.40673D-03	0.54225D-03	0.75018D 00
165.	0.40866D-03	0.43798D-03	0.93308D 00
180.	0.40931D-03	0.40931D-03	0.10000D 01

ALPHA = 0.04

THETA	RHO S	RHO P	RHO U
0.	0.30385D-03	0.30385D-03	0.10000D 01
5.	0.30418D-03	0.30644D-03	0.99262D 00
10.	0.30518D-03	0.31439D-03	0.97069D 00
20.	0.30912D-03	0.34887D-03	0.88608D 00
30.	0.31554D-03	0.41752D-03	0.75582D 00
40.	0.32424D-03	0.54513D-03	0.59492D 00
45.	0.32935D-03	0.64755D-03	0.50878D 00
50.	0.33493D-03	0.79359D-03	0.42224D 00
60.	0.34728D-03	0.13454D-02	0.25838D 00
70.	0.36090D-03	0.29356D-02	0.12326D 00
80.	0.37535D-03	0.11288D-01	0.33616D-01
90.	0.39020D-03	0.99067D 00	0.78376D-03
100.	0.40497D-03	0.14119D-01	0.29077D-01
110.	0.41924D-03	0.36529D-02	0.11514D 00
120.	0.43256D-03	0.17441D-02	0.24834D 00
130.	0.44456D-03	0.10792D-02	0.41220D 00
135.	0.44995D-03	0.90138D-03	0.49940D 00
150.	0.46322D-03	0.61762D-03	0.75012D 00
165.	0.47153D-03	0.50533D-03	0.93314D 00
180.	0.47436D-03	0.47436D-03	0.10000D 01

ALPHA = 0.06

THETA	RHO S	RHO P	RHO U
0.	0.19423D-03	0.19423D-03	0.10000D 01
5.	0.19496D-03	0.19636D-03	0.99289D 00
10.	0.19715D-03	0.20289D-03	0.97175D 00
20.	0.20585D-03	0.23132D-03	0.88991D 00
30.	0.22008D-03	0.28842D-03	0.76311D 00
40.	0.23948D-03	0.39585D-03	0.60508D 00
45.	0.25097D-03	0.48295D-03	0.51978D 00
50.	0.26354D-03	0.60805D-03	0.43358D 00
60.	0.29161D-03	0.10860D-02	0.26874D 00
70.	0.32292D-03	0.24746D-02	0.13077D 00
80.	0.35658D-03	0.96488D-02	0.37299D-01
90.	0.39164D-03	0.97818D 00	0.79170D-03
100.	0.42705D-03	0.16358D-01	0.26522D-01
110.	0.46174D-03	0.41537D-02	0.11157D 00
120.	0.49461D-03	0.20207D-02	0.24515D 00
130.	0.52460D-03	0.12800D-02	0.41014D 00
135.	0.53820D-03	0.10814D-02	0.49797D 00
150.	0.57205D-03	0.76286D-03	0.75002D 00
165.	0.59351D-03	0.63598D-03	0.93326D 00
180.	0.60087D-03	0.60087D-03	0.10000D 01

ALPHA = 0.08

THETA	RHO S	RHO P	RHO U
0.	0.34893D-04	0.34893D-04	0.10000D 01
5.	0.36149D-04	0.36394D-04	0.99328D 00
10.	0.39913D-04	0.41009D-04	0.97328D 00
20.	0.54889D-04	0.61294D-04	0.89551D 00
30.	0.79573D-04	0.10283D-03	0.77385D 00
40.	0.11354D-03	0.18309D-03	0.62017D 00
45.	0.13382D-03	0.24959D-03	0.53623D 00
50.	0.15618D-03	0.34666D-03	0.45063D 00
60.	0.20669D-03	0.72675D-03	0.28455D 00
70.	0.26404D-03	0.18569D-02	0.14242D 00
80.	0.32697D-03	0.76399D-02	0.43111D-01
90.	0.39396D-03	0.95894D 00	0.80446D-03
100.	0.46321D-03	0.21166D-01	0.22337D-01
110.	0.53270D-03	0.50665D-02	0.10562D 00
120.	0.60015D-03	0.25085D-02	0.23971D 00
130.	0.66312D-03	0.16326D-02	0.40657D 00
135.	0.69214D-03	0.13979D-02	0.49546D 00
150.	0.76569D-03	0.10214D-02	0.74983D 00
165.	0.81335D-03	0.87136D-03	0.93347D 00
180.	0.82986D-03	0.82986D-03	0.10000D 01

ALPHA = 0.02

THETA	RHO S	RHO P	RHO U
0.	0.14064D-02	0.14064D-02	0.10000D 01
5.	0.14066D-02	0.14172D-02	0.99248D 00
10.	0.14071D-02	0.14505D-02	0.97015D 00
20.	0.14093D-02	0.15943D-02	0.88413D 00
30.	0.14128D-02	0.18791D-02	0.75217D 00
40.	0.14175D-02	0.24048D-02	0.59002D 00
45.	0.14202D-02	0.28241D-02	0.50360D 00
50.	0.14232D-02	0.34191D-02	0.41708D 00
60.	0.14297D-02	0.56488D-02	0.25417D 00
70.	0.14368D-02	0.12003D-01	0.12097D 00
80.	0.14441D-02	0.44764D-01	0.33657D-01
90.	0.14516D-02	0.99772D 00	0.29022D-02
100.	0.14588D-02	0.46923D-01	0.32500D-01
110.	0.14656D-02	0.12463D-01	0.11889D 00
120.	0.14719D-02	0.58759D-02	0.25160D 00
130.	0.14774D-02	0.35718D-02	0.41450D 00
135.	0.14799D-02	0.29571D-02	0.50118D 00
150.	0.14858D-02	0.19802D-02	0.75069D 00
165.	0.14895D-02	0.15962D-02	0.93322D 00
180.	0.14907D-02	0.14907D-02	0.10000D 01

ALPHA = 0.04

THETA	RHO S	RHO P	RHO U
0.	0.12727D-02	0.12727D-02	0.10000D 01
5.	0.12734D-02	0.12829D-02	0.99264D 00
10.	0.12756D-02	0.13140D-02	0.97079D 00
20.	0.12842D-02	0.14489D-02	0.88645D 00
30.	0.12981D-02	0.17165D-02	0.75658D 00
40.	0.13169D-02	0.22111D-02	0.59612D 00
45.	0.13279D-02	0.26060D-02	0.51019D 00
50.	0.13398D-02	0.31668D-02	0.42385D 00
60.	0.13660D-02	0.52682D-02	0.26031D 00
70.	0.13947D-02	0.11232D-01	0.12539D 00
80.	0.14247D-02	0.41369D-01	0.35813D-01
90.	0.14552D-02	0.99070D 00	0.29198D-02
100.	0.14852D-02	0.50160D-01	0.31048D-01
110.	0.15137D-02	0.13103D-01	0.11687D 00
120.	0.15401D-02	0.61945D-02	0.24977D 00
130.	0.15635D-02	0.37912D-02	0.41331D 00
135.	0.15739D-02	0.31506D-02	0.50033D 00
150.	0.15993D-02	0.21319D-02	0.75059D 00
165.	0.16150D-02	0.17307D-02	0.93327D 00
180.	0.16204D-02	0.16204D-02	0.10000D 01

ALPHA = 0.06

THETA	RHO S	RHO P	RHO U
0.	0.10445D-02	0.10445D-02	0.10000D 01
5.	0.10461D-02	0.10536D-02	0.99293D 00
10.	0.10509D-02	0.10813D-02	0.97189D 00
20.	0.10699D-02	0.12017D-02	0.89044D 00
30.	0.11009D-02	0.14411D-02	0.76417D 00
40.	0.11428D-02	0.18851D-02	0.60669D 00
45.	0.11675D-02	0.22406D-02	0.52164D 00
50.	0.11945D-02	0.27460D-02	0.43565D 00
60.	0.12541D-02	0.46415D-02	0.27111D 00
70.	0.13200D-02	0.99928D-02	0.13324D 00
80.	0.13899D-02	0.36250D-01	0.39677D-01
90.	0.14618D-02	0.97830D 00	0.29518D-02
100.	0.15335D-02	0.57008D-01	0.28389D-01
110.	0.16027D-02	0.14349D-01	0.11312D 00
120.	0.16674D-02	0.68027D-02	0.24637D 00
130.	0.17257D-02	0.42086D-02	0.41105D 00
135.	0.17519D-02	0.35189D-02	0.49873D 00
150.	0.18164D-02	0.24221D-02	0.75039D 00
165.	0.18568D-02	0.19897D-02	0.93336D 00
180.	0.18706D-02	0.18706D-02	0.10000D 01

ALPHA = 0.08

THETA	RHO S	RHO P	RHO U
0.	0.71307D-03	0.71307D-03	0.10000D 01
5.	0.71584D-03	0.72065D-03	0.99333D 00
10.	0.72412D-03	0.74386D-03	0.97348D 00
20.	0.75699D-03	0.84469D-03	0.89626D 00
30.	0.81093D-03	0.10462D-02	0.77532D 00
40.	0.88466D-03	0.14222D-02	0.62236D 00
45.	0.92843D-03	0.17247D-02	0.53873D 00
50.	0.97644D-03	0.21563D-02	0.45336D 00
60.	0.10840D-02	0.37801D-02	0.28754D 00
70.	0.12047D-02	0.83471D-02	0.14535D 00
80.	0.13353D-02	0.30029D-01	0.45742D-01
90.	0.14724D-02	0.95936D 00	0.30028D-02
100.	0.16120D-02	0.71667D-01	0.24067D-01
110.	0.17500D-02	0.16611D-01	0.10691D 00
120.	0.18819D-02	0.78676D-02	0.24063D 00
130.	0.20034D-02	0.49343D-02	0.40720D 00
135.	0.20588D-02	0.41599D-02	0.49597D 00
150.	0.21978D-02	0.29324D-02	0.75005D 00
165.	0.22868D-02	0.24500D-02	0.93353D 00
180.	0.23174D-02	0.23174D-02	0.10000D 01

ALPHA = 0.02

THETA	RHO S	RHO P	RHO U
0.	0.49574D-02	0.49574D-02	0.10000D 01
5.	0.49578D-02	0.49952D-02	0.99254D 00
10.	0.49588D-02	0.51110D-02	0.97037D 00
20.	0.49630D-02	0.56116D-02	0.88499D 00
30.	0.49696D-02	0.66017D-02	0.75400D 00
40.	0.49784D-02	0.84237D-02	0.59303D 00
45.	0.49835D-02	0.98726D-02	0.50724D 00
50.	0.49890D-02	0.11923D-01	0.42134D 00
60.	0.50009D-02	0.19544D-01	0.25958D 00
70.	0.50135D-02	0.40783D-01	0.12731D 00
80.	0.50263D-02	0.14051D 00	0.40595D-01
90.	0.50389D-02	0.99772D 00	0.10039D-01
100.	0.50508D-02	0.14613D 00	0.39415D-01
110.	0.50617D-02	0.41914D-01	0.12519D 00
120.	0.50713D-02	0.20030D-01	0.25696D 00
130.	0.50796D-02	0.12218D-01	0.41870D 00
135.	0.50832D-02	0.10121D-01	0.50477D 00
150.	0.50916D-02	0.67776D-02	0.75249D 00
165.	0.50966D-02	0.54604D-02	0.93371D 00
180.	0.50982D-02	0.50982D-02	0.10000D 01

ALPHA = 0.04

THETA	RHO S	RHO P	RHO U
0.	0.47272D-02	0.47272D-02	0.10000D 01
5.	0.47286D-02	0.47635D-02	0.99271D 00
10.	0.47327D-02	0.48746D-02	0.97103D 00
20.	0.47491D-02	0.53550D-02	0.88739D 00
30.	0.47756D-02	0.63051D-02	0.75856D 00
40.	0.48108D-02	0.80527D-02	0.59934D 00
45.	0.48312D-02	0.94415D-02	0.51405D 00
50.	0.48532D-02	0.11404D-01	0.42834D 00
60.	0.49010D-02	0.18682D-01	0.26593D 00
70.	0.49520D-02	0.38813D-01	0.13188D 00
80.	0.50042D-02	0.13149D 00	0.42847D-01
90.	0.50558D-02	0.99074D 00	0.10108D-01
100.	0.51051D-02	0.15429D 00	0.37997D-01
110.	0.51505D-02	0.43393D-01	0.12321D 00
120.	0.51910D-02	0.20654D-01	0.25520D 00
130.	0.52259D-02	0.12606D-01	0.41760D 00
135.	0.52410D-02	0.10452D-01	0.50401D 00
150.	0.52770D-02	0.70251D-02	0.75246D 00
165.	0.52985D-02	0.56764D-02	0.93378D 00
180.	0.53056D-02	0.53056D-02	0.10000D 01

ALPHA = 0.06

THETA	RHO S	RHO P	RHO U
0.	0.43355D-02	0.43355D-02	0.10000D 01
5.	0.43386D-02	0.43693D-02	0.99300D 00
10.	0.43478D-02	0.44729D-02	0.97216D 00
20.	0.43842D-02	0.49204D-02	0.89150D 00
30.	0.44431D-02	0.58054D-02	0.76638D 00
40.	0.45221D-02	0.74321D-02	0.61022D 00
45.	0.45681D-02	0.87232D-02	0.52583D 00
50.	0.46178D-02	0.10546D-01	0.44047D 00
60.	0.47264D-02	0.17274D-01	0.27703D 00
70.	0.48436D-02	0.35670D-01	0.13996D 00
80.	0.49650D-02	0.11787D 00	0.46855D-01
90.	0.50864D-02	0.97853D 00	0.10232D-01
100.	0.52036D-02	0.17114D 00	0.35424D-01
110.	0.53133D-02	0.46243D-01	0.11958D 00
120.	0.54125D-02	0.21833D-01	0.25195D 00
130.	0.54989D-02	0.13337D-01	0.41553D 00
135.	0.55368D-02	0.11077D-01	0.50260D 00
150.	0.56277D-02	0.74936D-02	0.75240D 00
165.	0.56828D-02	0.60874D-02	0.93392D 00
180.	0.57012D-02	0.57012D-02	0.10000D 01

ALPHA = 0.08

THETA	RHO S	RHO P	RHO U
0.	0.37700D-02	0.37700D-02	0.10000D 01
5.	0.37753D-02	0.38004D-02	0.99342D 00
10.	0.37913D-02	0.38937D-02	0.97380D 00
20.	0.38546D-02	0.42969D-02	0.89747D 00
30.	0.39577D-02	0.50941D-02	0.77780D 00
40.	0.40969D-02	0.65580D-02	0.62624D 00
45.	0.41785D-02	0.77185D-02	0.54328D 00
50.	0.42674D-02	0.93535D-02	0.45854D 00
60.	0.44633D-02	0.15358D-01	0.29376D 00
70.	0.46779D-02	0.31537D-01	0.15230D 00
80.	0.49041D-02	0.10126D 00	0.53077D-01
90.	0.51345D-02	0.96016D 00	0.10429D-01
100.	0.53617D-02	0.20540D 00	0.31298D-01
110.	0.55788D-02	0.51318D-01	0.11366D 00
120.	0.57795D-02	0.23860D-01	0.24658D 00
130.	0.59581D-02	0.14583D-01	0.41206D 00
135.	0.60377D-02	0.12144D-01	0.50020D 00
150.	0.62318D-02	0.83008D-02	0.75229D 00
165.	0.63520D-02	0.68028D-02	0.93415D 00
180.	0.63926D-02	0.63926D-02	0.10000D 01

ALPHA = 0.02

THETA	RHO S	RHO P	RHO U
0.	0.97454D-02	0.97454D-02	0.10000D 01
5.	0.97459D-02	0.98192D-02	0.99261D 00
10.	0.97473D-02	0.10045D-01	0.97066D 00
20.	0.97529D-02	0.11020D-01	0.88611D 00
30.	0.97617D-02	0.12946D-01	0.75641D 00
40.	0.97734D-02	0.16480D-01	0.59699D 00
45.	0.97801D-02	0.19281D-01	0.51203D 00
50.	0.97872D-02	0.23229D-01	0.42695D 00
60.	0.98024D-02	0.37768D-01	0.26673D 00
70.	0.98181D-02	0.77174D-01	0.13571D 00
80.	0.98337D-02	0.24304D 00	0.49806D-01
90.	0.98485D-02	0.99773D 00	0.19527D-01
100.	0.98620D-02	0.25136D 00	0.48616D-01
110.	0.98738D-02	0.78977D-01	0.13358D 00
120.	0.98837D-02	0.38486D-01	0.26409D 00
130.	0.98917D-02	0.23630D-01	0.42429D 00
135.	0.98950D-02	0.19606D-01	0.50954D 00
150.	0.99024D-02	0.13160D-01	0.75488D 00
165.	0.99065D-02	0.10610D-01	0.93435D 00
180.	0.99078D-02	0.99078D-02	0.10000D 01

ALPHA = 0.04

THETA	RHO S	RHO P	RHO U
0.	0.94730D-02	0.94730D-02	0.10000D 01
5.	0.94749D-02	0.95444D-02	0.99278D 00
10.	0.94805D-02	0.97630D-02	0.97134D 00
20.	0.95026D-02	0.10707D-01	0.88857D 00
30.	0.95379D-02	0.12570D-01	0.76107D 00
40.	0.95846D-02	0.15984D-01	0.60344D 00
45.	0.96113D-02	0.18686D-01	0.51899D 00
50.	0.96399D-02	0.22489D-01	0.43410D 00
60.	0.97010D-02	0.36445D-01	0.27323D 00
70.	0.97647D-02	0.73967D-01	0.14041D 00
80.	0.98282D-02	0.22943D 00	0.52153D-01
90.	0.98888D-02	0.99078D 00	0.19675D-01
100.	0.99443D-02	0.26305D 00	0.47279D-01
110.	0.99934D-02	0.81239D-01	0.13169D 00
120.	0.10035D-01	0.39346D-01	0.26244D 00
130.	0.10069D-01	0.24117D-01	0.42331D 00
135.	0.10083D-01	0.20008D-01	0.50890D 00
150.	0.10115D-01	0.13442D-01	0.75493D 00
165.	0.10132D-01	0.10851D-01	0.93445D 00
180.	0.10138D-01	0.10138D-01	0.10000D 01

ALPHA = 0.06

THETA	RHO S	RHO P	RHO U
0.	0.90119D-02	0.90119D-02	0.10000D 01
5.	0.90161D-02	0.90795D-02	0.99308D 00
10.	0.90286D-02	0.92863D-02	0.97249D 00
20.	0.90776D-02	0.10179D-01	0.89275D 00
30.	0.91565D-02	0.11940D-01	0.76902D 00
40.	0.92610D-02	0.15159D-01	0.61450D 00
45.	0.93213D-02	0.17701D-01	0.53097D 00
50.	0.93858D-02	0.21271D-01	0.44644D 00
60.	0.95247D-02	0.34298D-01	0.28452D 00
70.	0.96711D-02	0.68878D-01	0.14864D 00
80.	0.98184D-02	0.20874D 00	0.56302D-01
90.	0.99606D-02	0.97875D 00	0.19939D-01
100.	0.10093D-01	0.28648D 00	0.44870D-01
110.	0.10211D-01	0.85548D-01	0.12826D 00
120.	0.10312D-01	0.40955D-01	0.25943D 00
130.	0.10396D-01	0.25023D-01	0.42148D 00
135.	0.10432D-01	0.20756D-01	0.50771D 00
150.	0.10512D-01	0.13970D-01	0.75502D 00
165.	0.10557D-01	0.11304D-01	0.93463D 00
180.	0.10572D-01	0.10572D-01	0.10000D 01

ALPHA = 0.08

THETA	RHO S	RHO P	RHO U
0.	0.83511D-02	0.83511D-02	0.10000D 01
5.	0.83584D-02	0.84135D-02	0.99350D 00
10.	0.83801D-02	0.86044D-02	0.97415D 00
20.	0.84657D-02	0.94282D-02	0.89878D 00
30.	0.86040D-02	0.11050D-01	0.78054D 00
40.	0.87886D-02	0.14008D-01	0.63066D 00
45.	0.88958D-02	0.16337D-01	0.54855D 00
50.	0.90112D-02	0.19597D-01	0.46464D 00
60.	0.92618D-02	0.31406D-01	0.30137D 00
70.	0.95295D-02	0.62242D-01	0.16110D 00
80.	0.98034D-02	0.18331D 00	0.62667D-01
90.	0.10073D-01	0.96091D 00	0.20350D-01
100.	0.10328D-01	0.33150D 00	0.41058D-01
110.	0.10560D-01	0.93057D-01	0.12275D 00
120.	0.10765D-01	0.43673D-01	0.25452D 00
130.	0.10938D-01	0.26542D-01	0.41847D 00
135.	0.11012D-01	0.22011D-01	0.50574D 00
150.	0.11184D-01	0.14864D-01	0.75517D 00
165.	0.11284D-01	0.12079D-01	0.93494D 00
180.	0.11317D-01	0.11317D-01	0.10000D 01

ALPHA = 0.02

THETA	RHO S	RHO P	RHO U
0.	0.15136D-01	0.15136D-01	0.10000D 01
5.	0.15136D-01	0.15249D-01	0.99269D 00
10.	0.15138D-01	0.15597D-01	0.97098D 00
20.	0.15144D-01	0.17100D-01	0.88736D 00
30.	0.15155D-01	0.20061D-01	0.75907D 00
40.	0.15168D-01	0.25478D-01	0.60139D 00
45.	0.15176D-01	0.29756D-01	0.51734D 00
50.	0.15184D-01	0.35762D-01	0.43318D 00
60.	0.15201D-01	0.57655D-01	0.27467D 00
70.	0.15217D-01	0.11526D 00	0.14504D 00
80.	0.15234D-01	0.33317D 00	0.60043D-01
90.	0.15248D-01	0.99773D 00	0.30073D-01
100.	0.15261D-01	0.34311D 00	0.58841D-01
110.	0.15271D-01	0.11765D 00	0.14289D 00
120.	0.15279D-01	0.58568D-01	0.27200D 00
130.	0.15285D-01	0.36238D-01	0.43049D 00
135.	0.15287D-01	0.30127D-01	0.51482D 00
150.	0.15291D-01	0.20284D-01	0.75753D 00
165.	0.15292D-01	0.16372D-01	0.93506D 00
180.	0.15293D-01	0.15293D-01	0.10000D 01

ALPHA = 0.04

THETA	RHO S	RHO P	RHO U
0.	0.14866D-01	0.14866D-01	0.10000D 01
5.	0.14868D-01	0.14976D-01	0.99287D 00
10.	0.14875D-01	0.15315D-01	0.97167D 00
20.	0.14900D-01	0.16775D-01	0.88987D 00
30.	0.14941D-01	0.19652D-01	0.76382D 00
40.	0.14995D-01	0.24905D-01	0.60797D 00
45.	0.15025D-01	0.29046D-01	0.52444D 00
50.	0.15057D-01	0.34851D-01	0.44048D 00
60.	0.15125D-01	0.55925D-01	0.28132D 00
70.	0.15194D-01	0.11094D 00	0.14988D 00
80.	0.15260D-01	0.31661D 00	0.62503D-01
90.	0.15320D-01	0.99082D 00	0.30317D-01
100.	0.15372D-01	0.35661D 00	0.57593D-01
110.	0.15415D-01	0.12057D 00	0.14109D 00
120.	0.15448D-01	0.59601D-01	0.27046D 00
130.	0.15472D-01	0.36769D-01	0.42960D 00
135.	0.15481D-01	0.30548D-01	0.51427D 00
150.	0.15498D-01	0.20557D-01	0.75764D 00
165.	0.15504D-01	0.16597D-01	0.93518D 00
180.	0.15506D-01	0.15506D-01	0.10000D 01

ALPHA = 0.06

THETA	RHO S	RHO P	RHO U
0.	0.14411D-01	0.14411D-01	0.10000D 01
5.	0.14416D-01	0.14517D-01	0.99317D 00
10.	0.14431D-01	0.14840D-01	0.97284D 00
20.	0.14488D-01	0.16232D-01	0.89410D 00
30.	0.14580D-01	0.18971D-01	0.77188D 00
40.	0.14700D-01	0.23958D-01	0.61917D 00
45.	0.14769D-01	0.27879D-01	0.53659D 00
50.	0.14841D-01	0.33360D-01	0.45299D 00
60.	0.14995D-01	0.53138D-01	0.29279D 00
70.	0.15152D-01	0.10412D 00	0.15829D 00
80.	0.15305D-01	0.29132D 00	0.66821D-01
90.	0.15447D-01	0.97897D 00	0.30750D-01
100.	0.15570D-01	0.38299D 00	0.55363D-01
110.	0.15674D-01	0.12605D 00	0.13786D 00
120.	0.15755D-01	0.61516D-01	0.26765D 00
130.	0.15814D-01	0.37750D-01	0.42797D 00
135.	0.15837D-01	0.31326D-01	0.51327D 00
150.	0.15881D-01	0.21063D-01	0.75784D 00
165.	0.15901D-01	0.17017D-01	0.93539D 00
180.	0.15906D-01	0.15906D-01	0.10000D 01

ALPHA = 0.08

THETA	RHO S	RHO P	RHO U
0.	0.13766D-01	0.13766D-01	0.10000D 01
5.	0.13774D-01	0.13864D-01	0.99359D 00
10.	0.13800D-01	0.14166D-01	0.97450D 00
20.	0.13900D-01	0.15466D-01	0.90016D 00
30.	0.14061D-01	0.18018D-01	0.78345D 00
40.	0.14274D-01	0.22650D-01	0.63540D 00
45.	0.14395D-01	0.26277D-01	0.55424D 00
50.	0.14525D-01	0.31332D-01	0.47128D 00
60.	0.14803D-01	0.49420D-01	0.30974D 00
70.	0.15091D-01	0.95294D-01	0.17087D 00
80.	0.15375D-01	0.26005D 00	0.73368D-01
90.	0.15642D-01	0.96163D 00	0.31416D-01
100.	0.15881D-01	0.43124D 00	0.51882D-01
110.	0.16084D-01	0.13541D 00	0.13273D 00
120.	0.16248D-01	0.64693D-01	0.26313D 00
130.	0.16373D-01	0.39367D-01	0.42531D 00
135.	0.16421D-01	0.32608D-01	0.51161D 00
150.	0.16520D-01	0.21905D-01	0.75817D 00
165.	0.16566D-01	0.17724D-01	0.93576D 00
180.	0.16579D-01	0.16579D-01	0.10000D 01